The BIBLE SPEAKS FOR ITSELF

GEORGE A. HALL

Wasteland Press
www.wastelandpress.net
Shelbyville, KY USA

The Bible Speaks For Itself
by George A. Hall

First Printing – September 2014
Paperback ISBN: 978-1-60047-991-5
The author can be contacted at *preachergeorgewv@yahoo.com.*

The English Standard Version of the Bible, 2011, is the text of choice for this work. Unless a citation specifies otherwise, all quotations used in this work are from the English Standard Version (Hereafter ESV).

Where **bold fonts** occur they show emphasis by this author; they were not emboldened in the original ESV.

Pictures are owned by this author unless otherwise stated.

Printed in the U.S.A.

0 1 2 3 4 5 6

TABLE OF CONTENTS

ACKNOWLEDGMENTS

Many students over the years have studied this material in some format or another. I thank each of them for bearing with the materials and my deficiencies over the time. The same forbearance is requested for this material as well. Those questions and suggestions from learners were often written into my marginal notes for any future revision. I hope this book reflects some of those suggestions with the same sensitivity in which they were given.

I am also very appreciative of those whose recent editorial contributions have made this work possible. My wife, children, other family members, and friends have been helpful and encouraging. I thank you very much!

INTRODUCTION

For lots of devout Christians the Bible is God's Word simply because it says so. That is an article of faith. The passage most often cited in this belief is 2 Timothy 3:16, translated as:

> *"All Scripture is given by inspiration of God."* King James Version
> *"All Scripture is God-breathed."* New International Version, 1984
> *"All Scripture is breathed out by God."* English Standard Version

But broader popular opinion has been shifting away from such a belief. Skepticism about the inspiration and the historical nature of the Bible has grown. Over several years there have been numerous cover articles in leading news magazines that challenge the basic facts. A growing multitude of people watch historical documentaries on the subject that are not friendly to the traditional view that the Bible is inspired by God. Even the film industry has joined, cashing in on conspiracy theories of one form or another, the best known being based on Dan Brown's novel *The Da Vinci Code.*

Common threads to the modern assault on the Bible include the assumption that the present Bible evolved over several generations. According to many, the New Testament is based only on oral traditions begun by the Apostles, and only written much later. Later yet, theological partisans of the church are said to have formed councils to ban alleged books of the Bible that they did not like (usually those that are Gnostic in orientation and theology).

What has not been clear in many modern accounts about the Bible is what the Bible itself actually tells us about its own historical background and inspiration. Several passages give insights into how the Scriptures came into being. It is with these lesser known passages in mind that this book has been titled: *The Bible Speaks for Itself.* This book will show how the Bible presents its own origin; then, those points will be compared with what passes for good history today.

A comparison of the history of the Bible will be made to the history of the Holocaust and Nazi Germany in chapter four. It appears that there will always be detractors of the Holocaust. The motives are various, and often religious, for denying the Holocaust. The same can be said about some denials of the Bible's historical integrity and inspiration. As in the case for the Holocaust, the existence of detractors of the Bible is not a clear reflection or measure of the quality of the evidence at hand. As long as there are people who hate the Jewish people there will be detractors of the historical reality of the Holocaust: those who want to discredit the Bible will do so. But let them please do so based on its own testimony of how it came into being. One may not agree with the message of the gospel, but that doesn't mean that the written sources lack actual historical integrity.

Too often the real reason that most people do not seek God or believe him has nothing to do with the true level of the evidence in God's Word. It has more to do with not knowing the Word, or rejecting it because we do not want to change. Chapters six and seven will deal with serious background issues that affect those who do not believe the testimony of the Bible. Appendix A is provided for those who might want to become a Christian as a result learning how the Bible speaks for itself on its origins and purpose.

Please proceed prayerfully and keep in mind that the cumulative weight of the Bible passages is important. Taken all together they present a case for divine inspiration that is consistent and helpful. The Bible does speak for itself in this matter.

CHAPTER 1:

The Inspiration of the Old Testament

"You shall not add to the word that I command you, nor take from it, that you may keep the commandments of the LORD your God that I command you."

Moses, Deuteronomy 4:2

An overview is in order before looking at the first five books of the Bible. The pictured Hebrew Bible is the one commonly used by scholars. It is full of notes and cross references about the nature of the various manuscript sources that are used in it. It is a very carefully documented work, used by translators and commentators alike.

Above are the cover and a data page of a scholarly Hebrew Bible.

Below is a portion of Genesis, the first book of the Old Testament, the Bible of the Jewish people. It is chapter 1:1-5 from the modern edition of the Hebrew Bible just shown. This language actually reads from right to left, as you might have guessed already.

GENESIS בראשית

[ס] 1 1 בראשית ברא אלהים את השמים ואת הארץ׃ 2 והארץ
היתה תהו ובהו וחשך על־פני תהום ורוח אלהים מרחפת על־פני
3 4 המים׃ 3 ויאמר אלהים יהי אור ויהי־אור׃ 4 וירא אלהים את־
5 האור כי־טוב ויבדל אלהים בין האור ובין החשך׃ 5 ויקרא

Foundational to the authority of the Christian Bible is the authority of the Jewish Bible, which is also the Old Testament portion of the Christian Bible. Christians add the New Testament to the Jewish Bible because they believe that Jesus was the Messiah promised in the Old Testament. Those who practice the Jewish religion therefore do not include any of the New Testament.

The writing of the Old Testament, as a body of authoritative truth from God, began with the writing of Moses. Several references in the first five books of the Old Testament, taken together, give the reader a glimpse into how these writings were and are believed to be inspired by God. Further, they give a clear picture of how the content of these books was to be viewed and preserved for the future of the faithful.

The age of these first five books can be known by a reference in First Kings, where we find the words

> *I Kings 6:1 In the four hundred and eightieth year after the people of Israel came out of the land of Egypt, in the fourth year of Solomon's reign over Israel, in the month of Ziv, which is the second month, he [Solomon] began to build the house of the LORD.*

Moses did his writing during the 40 years after the exodus from Egypt. The exodus is said to have happened 480 years before Solomon began to rule in 967 BC (jewishvirtuallibrary.org/jsource/biography/Solomon). If the noted 480 years is added to Solomon's reign at 967 BC the exodus can be dated to the year 1447 BC. That means that fourteen and a half centuries before Christ,

God began to compile his will in written form. This written record would establish God's will for the nation of Israel.

Moses Had Authority From God

Chapter three of Exodus tells us that Moses was attracted to a burning bush that was not consumed by the flames. What a way to get the attention of Moses because brush fires are greatly feared in arid areas. They spread quickly, destroying everything, including flocks and herds. Moses probably rushed there to put it out. But from within that bush God began to speak his will to Moses. God told Moses how he was to convey these expectations to others by saying such things as found in these citations:

> Exodus 4:12 *Now therefore go, and I will be with your mouth and teach you what you shall speak." 13 But he said, 'Oh, my Lord, please send someone else.' 14 Then the anger of the LORD was kindled against Moses and he said, "Is there not Aaron, your brother, the Levite? I know that he can speak well. Behold, he is coming out to meet you, and when he sees you, he will be glad in his heart. 15 You shall speak to him and put the words in his mouth, and I will be with your mouth and with his mouth and will teach you both what to do.*

> Exodus 35:1 *Moses assembled all the congregation of the people of Israel and said to them, "These are the things that the LORD has commanded you to do.*

God used Moses over a period of forty years in the desert to turn a mob of slaves from Egypt into an organized nation of people. He taught them in detail to be citizens of a nation that worshipped the one true God. He used laws and symbolism that would teach them about his holiness. God made it clear that Moses was speaking for God and not himself. Moses obeyed God in what was to be done with those laws: they were to be written.

God Emphasized a Permanent Written Record

The following citations reflect the importance God placed on a permanent written record:

> Exodus 17:14 *Then the LORD said to Moses, '**Write this as a memorial in a book** and recite it in the ears of Joshua ....'* [Bold font is always this author's emphasis in the ESV translation.]

> Exodus 24:3-4 *Moses came and told the people **all the words** of the LORD and all the rules. And **all** the people answered with one voice and said, '**All the words** that the LORD has spoken we will do.' 4 And **Moses wrote down all the words** of the Lord.*

> Exodus34:1, 4 *The LORD said to Moses, 'Cut for yourself two tablets of stone like the first, and **I will write on the tablets the words that were on the first tablets**, which you broke. ... 4 So Moses cut two tablets of stone like the first. And he rose early in the morning and went up on Mount Sinai, as the LORD had commanded him, and took in his hand two tablets of stone.*

Not only was it important to God that his Word be recorded permanently, but also he stressed the importance of never changing it in the least. Those entrusted with God's Word were never to add to it and never to take away from it. It was not only important to write it, but to record it accurately for the future.

God Wanted No Changes in His Word

His people were to be very careful to preserve God's Word, just as he had given it through Moses. Such fixed teachings would be necessary for a faith that would have to be passed down accurately, over many generations. Some references that reflect this expectation for preserving the Word of God are:

> Deuteronomy 4:1 *And now, O Israel, listen to the statutes and the rules that I am teaching you, and do them, that you may live, and go in and take possession of the land that the LORD, the God of your fathers, is giving you. 2 You shall **not add to the word that I command you, nor take from it**, that you may keep the commandments of the LORD your God that I command you.*

> Deuteronomy 12:32 *Everything that I command you, you shall be careful to do. You shall **not add to it or take from it**.*

> Deuteronomy 31:24 *When Moses had finished **writing the words of this law in a book to the very end**, 25 Moses commanded the*

> *Levites who carried the ark of the covenant of the LORD, 26 "Take this Book of the Law and* ***put it by the side of the ark*** *of the covenant of the LORD your God, that it may be there* ***for a witness against you****.*

In these citations we see a concern to insure the future accuracy of the text of God's Word. An original was written by Moses himself and placed by the ark. It would be a witness to its content and to their enthusiastic agreement as a nation to obey it. The next generation reinforced this agreement when they entered the Promised Land.

Future Generations to be Taught Diligently

After the time of Moses, scribes were known for being careful to record all copies as accurately as possible. These Jewish religious leaders were respected as those who transcribed the Word of God into the many copies needed in the nation. Some of the passages related to passing on God's Word to future generations follow.

> Deuteronomy 6:4 *Hear, O Israel: The LORD our God, the LORD is one. 5 You shall love the LORD your God with all your heart and with all your soul and with all your might. 6 And* ***these words that I command you*** *today shall be on your heart. 7 You shall* ***teach them diligently*** *to your children, and shall* ***talk of them*** *when you sit in your house, and when you walk by the way, and when you lie down, and when you rise. 8 You shall* ***bind them*** *as a sign on your hand, and they shall be as frontlets between your eyes. 9 You shall* ***write them*** *on the doorposts of your house and on your gates.*

> Deuteronomy 32:45 *And when Moses had finished speaking* ***all these words*** *to all Israel, 46 he said to them, "Take to heart all the words by which I am warning you today, that you may* ***command them to your children****, that they may be careful to do* ***all the words*** *of this law. 47 For it is no empty word for you, but your very life, and by this word you shall live long in the land that you are going over the Jordan to possess.*

> Joshua 8:30 *At that time Joshua built an altar to the LORD, the God of Israel, on Mount Ebal, 31 just as Moses the servant of the LORD had commanded the people of Israel, as it is written in the Book of the Law of Moses, "an altar of uncut stones, upon which*

> *no man has wielded an iron tool." And they offered on it burnt offerings to the LORD and sacrificed peace offerings 32 And there, in the presence of the people of Israel,* ***he wrote on the stones a copy of the law of Moses, which he had written****. 33 And all Israel, sojourner as well as native born, with their elders and officers and their judges, stood on opposite sides of the ark before the Levitical priests who carried the ark of the covenant of the LORD, half of them in front of Mount Gerizim and half of them in front of Mount Ebal, just as Moses the servant of the LORD had commanded at the first, to bless the people of Israel. 34 And afterward he read all the words of the law, the blessing and the curse, according to* ***all that is written*** *in the Book of the Law. 35 There was not a word of all that Moses commanded that Joshua did not read before all the assembly of Israel, and the women, and the little ones, and the sojourners who lived among them.*

For God's Word to be passed through the generations required accurate repetition.

In summary, Moses is held to be the inspired leader of the fledgling nation of Israel. He wrote the first five books of the Old Testament. He was led by God to write in a way that Scripture would be preserved till the Messiah would come. So the writing of the books of Moses was commanded by God. Those Scriptures were recognized by Jesus as God's Word, and those Scriptures were said by Jesus to "bear witness about me," and that "if you believed Moses, you would believe me, for he wrote of me" (John 5:46-47).

But, what about those prophetic writings that came after the time of Moses?

Prophets and Prophecy in the Old Testament

The prophetic books themselves make it clear that they were believed to be the very words of God to his people. The statements were clear and frequent about being from God to his people. For example, in major prophetic books, such as Isaiah and Ezekiel, their statements were cited, at times, as coming from visions. In some cases, a clear commission by God is conveyed within the vision. In all cases, the historical context and dates of the words are noted for the readers. For example, in Ezekiel we learn how there is an immediate context (1:1) and a reference to inspiration (2:1):

> Ezekiel 1:1 *In the thirtieth year, in the fourth month, on the fifth day of the month, as I was among the exiles by the Chebar canal, the heavens were opened, and I saw* ***visions of God****.*

> Ezekiel 2:1 *And he said to me, "Son of man, stand on your feet, and I will speak with you." 2 And as he spoke to me,* ***the Spirit entered into me*** *and set me on my feet, and I heard him speaking to me. 3 And he said to me, "Son of man, I send you to the people of Israel, to nations of rebels, who have rebelled against me. They and their fathers have transgressed against me to this very day.*

In the smaller and more minor prophetic books, the wording is similar to what is found here in Haggai:

> Haggai 1:1 *In the second year of Darius the king, in the sixth month, on the first day of the month, the* ***word of the LORD*** *came by the hand of Haggai the prophet to Zerubbabel the son of Shealtiel, governor of Judah, and to Joshua the son of Jehozadak, the high priest: 2 "Thus says the LORD of hosts: These people say the time has not yet come to rebuild the house of the LORD." 3 Then* ***the word of the LORD*** *came by the hand of Haggai the prophet,…*

Divine Interpretation of Old Testament Prophecy

The futuristic aspect of God's revelation in the Old Testament is noted in a key passage from the Apostle Peter. This future time was a subject of curiosity, even by those under the Old Covenant.

> *I Peter 1:10 Concerning this salvation, the prophets who prophesied about the grace that was to be yours searched and inquired carefully, 11 inquiring what person or time* ***the Spirit of Christ in them*** *was indicating when* ***he predicted*** *the sufferings of Christ and the subsequent glories. 12 It was revealed to them that* ***they were serving not themselves but you****, in the things that have now been announced to you through those who preached the good news to you by the Holy Spirit sent from heaven,* ***things into which angels long to look****. … 22 Having purified your souls by your obedience to the truth for a sincere brotherly love, love one another earnestly from a pure heart, 23 since you have been born again, not of perishable seed but of imperishable, through the living and abiding word of God; 24 for 'All flesh is like grass and all its glory like the flower of grass. The grass withers, and the flower falls,*

> *25 but the word of the Lord remains forever. 'And* ***this word is the good news*** *that was preached to you.*

From this passage we also learn that the Spirit of God was at work under the Old Covenant (v.11) and the New Covenant (v.12). Believers are born again by this imperishable, living, and abiding Word of God (vv.22-23). Peter emphasizes the imperishable quality by quoting Isaiah 40:6-8. He then adds that "this word is the good news that was preached to you" (v.25).

The second letter of Peter goes farther than his first letter when it comes to the subject of Scriptures and inspiration. In his second letter, Peter says:

> *2 Peter 1:16 For we did* ***not*** *follow* ***cleverly devised myths*** *when we made known to you the power and coming of our Lord Jesus Christ, but we were* ***eyewitnesses*** *of his majesty. 17 For when he received honor and glory from God the Father, and the voice was borne to him by the Majestic Glory, "This is my beloved Son, with whom I am well pleased," 18* ***we ourselves heard this very voice*** *borne from heaven, for* ***we were with him*** *on the holy mountain. 19 And we have* ***the prophetic word more fully confirmed,*** *to which you will do well to pay attention as to a lamp shining in a dark place,... knowing* ***this first of all,*** *that no prophecy of Scripture comes from someone's own interpretation. 21 For no prophecy was ever produced by the will of man, but* ***men spoke from God as they were carried along by the Holy Spirit.***

Peter witnessed and experienced a real Jesus. His expression in verse 16, that "we did not follow cleverly devised myths," has to have application at least to the fact that Peter's is not a mythical portrayal of a Jesus created by the memory of generations of wishful thinkers. Notice also, in the wording of 2 Peter 1:16, how he shares eyewitness accounts. He says that "we" were eyewitnesses, meaning the Apostolic band and others who saw the same things. He specifically includes personal incidents from the gospel accounts, even to the fact of hearing the voice from heaven say, "this is my beloved Son, with whom I am well pleased."

He then concludes with the necessary implication, "we have the prophetic word more fully confirmed, to which you will do well to pay attention," meaning the Old Testament prophesies are now confirmed in the life, deeds, and words of Jesus. Of defining importance for this point are the words that are so emphatic in these verses:

> *2 Peter 1:20 ...**no prophecy of Scripture comes from someone's own interpretation**. 21 For no prophecy was ever produced by the will of man, but **men spoke from God as they were carried along by the Holy Spirit**.*

The Old Testament prophetic books were historically defined by the immediate setting to which they spoke. They primarily existed to call the people back to their original covenant with God; however, some divine utterances referred to a future Messianic coming.

The Apocrypha of the Roman Catholic Version of the Bible

A centuries-old disagreement exists between Roman Catholic and Protestant Bible translators. The issue is the status of seven Apocryphal (meaning of spurious or doubtful origin) books attributed to a Greek translation of the Old Testament. Completed by around 130 B. C., this Greek translation, called the Septuagint, saw fit to include writings that were of historical interest to the Jewish faith. Presumably they were added to aid the understanding of Jewish history by those whose main language was Greek. They were never part of any original Hebrew edition of the Jewish Bible. The Hebrew editions of the Bible all stopped with the prophet Malachi.

This adding of material would likely have been to them much like what happens when modern publishers issue a study Bible. Additional material is presented for the benefit of the reader, but it is not to be confused with the actual text itself. Given enough time this assumed distinction was lost for the average reader of the Septuagint and even church leaders. Five centuries later the average Christians who used the Septuagint were accustomed to the presence of those extra books. In fact, some church leaders like the renowned Augustine viewed the complete Septuagint itself as inspired by God. If Augustine believed in the inspiration of the Septuagint, and hence the Apocrypha, one assumes that the same belief was widespread among Christians.

By the 390s A. D., a scholar of the Catholic Church started work on a major Latin translation of the Bible, called the Vulgate. It was completed by Jerome in 405 A. D. His translation was based on the Hebrew texts alone and not the Greek Septuagint. Yet, Jerome retained the Apocryphal books because of the reverence with which the Septuagint was held. To be precise about the distinction, he was clear that the extra books were classified as Apocryphal.

Further, he made various other notes on the texts such as book introductions. One example of how he held the distinction comes from the preface to the books of Samuel:

> *"This preface to the Scriptures may serve as a helmeted [i.e. defensive] introduction to all the books which we turn from Hebrew into Latin, so that we may be assured that what is outside of them must be placed aside among the Apocryphal writings."* Brackets in source article (www.ask.com/wiki/Jerome).

This inclusion of the Apocryphal books in the Vulgate became a major issue over a thousand years later during the Protestant Reformation. Protestant translators faced the decision of whether or not to preserve the Apocrypha which amounted to a major addition to the Word of God. They chose not to include it because adding to God's Word was forbidden.

Not only were the Apocryphal books not in the authoritative Jewish Bibles, they lacked the clear character and claims to inspiration that existed in the regular inspired texts of the Jewish Bible. Thirdly, I Maccabees actually said that the divine inspiration of the Hebrew Bible had stopped (I Maccabees 4:46; 9:27; 14:41). Fourthly, there are no future prophetic references to the Messiah. Fifthly, various statements in those books actually contradict sound teaching contained in the books that were clearly regarded as inspired. Keeping such Apocryphal books, even for "reader value," was not supported by Protestant translators. The easy answer for Reformers and Restorers of the church was to omit books that did not belong there in the first place.

Ironically, this decision was not made immediately or comprehensively. The earliest versions of the English King James Bible, also called the Authorized Version (first approved in 1611), included the Apocryphal books. In fact, the Bible pictured at the end of chapter two has the Apocrypha in the table of contents as late as 1766, but the books themselves were omitted from that same text. The result is that most Bibles sold in a store today will only have the regular 39 books of the Old Testament, and then the 27 books of the New Testament.

Summary

This chapter has shown how God used Moses to deliver his Word. Recorded for us in those words from God is how careful Moses and the people were to be in preserving God's word. Much is said about the accurate copying of God's written word, of not adding to it, nor taking away from it. God's word is to be conveyed to each generation in exact detail. After the death of Moses, God used many prophets to call the people back to himself, to record what God had told them, including more and more details of a coming Messiah who would save them from their sins.

The present 39 books of the Christian Old Testament are the same books that make up the original Jewish Bible. The content of this Scripture was to be known by the kings and priests and taught faithfully to all the people. All were to know the holiness and love of God. They were also to be anticipating the coming of the Messiah who would deal with their sins.

This same concern for accuracy was also shown by God in the New Testament when he spoke through his Apostles and others whose writings make up the New Testament. Let us now turn to those writings.

> *"Fragments of every book of the Hebrew Bible except for the book of Esther have been found in the Dead Sea Scroll collection found at Qumran. Some of these copies are identical to the traditional text of the Hebrew Bible used today."* www.deadseascrolls.org.il/featured-scrolls

Discussion

1. Can you state a simple way to show the importance of accurate written accounts in the Old Testament?

2. What five events do most people know from the Old Testament? How do those stories fit into the idea of a Messiah to save us from our sins?

3. Which part of the book of Exodus is the most important to a coming Messiah, the Ten Commandments (chapter 20) , or the details of the Passover (chapters 12 and 13)?

4. How has the discovery of the Dead Sea Scrolls enhanced the integrity of the Old Testament as historically preserved?

5. What parts of the Old Testament are the most important to emphasize to enhance our Bible literacy?

6. What have you found to be the most fascinating in studying this chapter?

CHAPTER 2:

The Inspiration of the New Testament

*"There are some things in them [Paul's letters] that are hard to understand, which the ignorant and unstable twist to their own destruction, as they do **the other Scriptures**."*

2 Peter 3:16

Pictured below is a Greek New Testament, turned to chapter 24 of the Gospel of Matthew. The language in which it is written is called Koine Greek. It is not classical Greek, but a more recent and common form of the language. It is a language not spoken today.

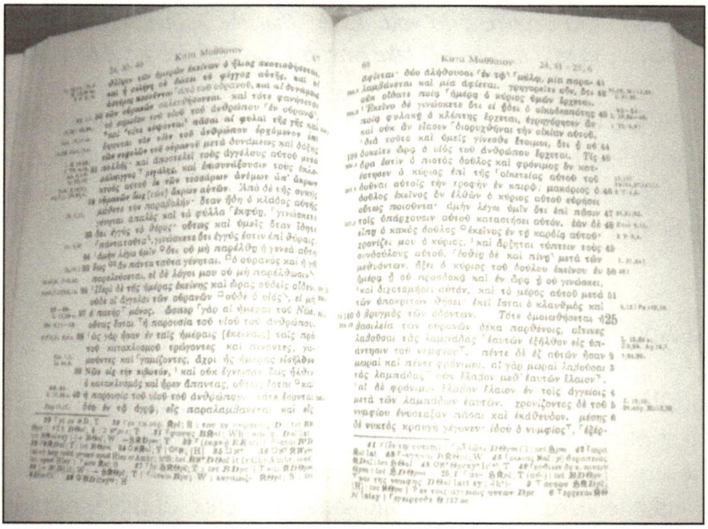

Jesus Said He Would Leave His Truth for All

The Scriptures were always to be preserved till the end of time, as Jesus stated in Matthew 5:18, "*For truly, I say to you, until heaven and earth pass away, not an iota, not a dot, will pass from the Law until all is accomplished.*"

In the course of correcting the theology of his critics, Jesus made a very clear statement about the authority of the Scriptures by saying, in John 10:35, that "Scripture cannot be broken," meaning disobeyed or discarded. That these declarations included more than the authority of the Old Testament can be seen from His words to his disciples in the Gospel of John.

> John 14:23 *Jesus answered him, "If anyone loves me, he will keep my word, and my Father will love him, and we will come to him and make our home with him. 24 Whoever does not love me does not keep my words. And the word that you hear is* ***not mine but the Father's*** *who sent me. 25 "These things I have spoken to you while I am still with you. 26 But the Helper, the Holy Spirit, whom the Father will send in my name,* ***he will teach you all things and bring to your remembrance all that I have said to you.***
>
> John 16:12 *I still have many things to say to you, but you cannot bear them now. 13 When the Spirit of truth comes, he will guide you into all the truth, for he will not speak on his own authority, but whatever he hears he will speak, and he will declare to you the things that are to come. 14 He will glorify me, for* ***he will take what is mine and declare it to you****. 15 All that the Father has is mine; therefore I said that* ***he will take what is mine and declare it to you.***

John 14:26 is clear in saying that the disciples would receive future inspiration by the Holy Spirit. John 16:12 says that the disciples could not bear the inspiration of the Holy Spirit right now, but it would happen in the future, after the Holy Spirit comes.

Inspired Apostles

Two Apostles who wrote and died before 67 A. D. have much to say about this subject of inspiration. God used them in the process of inspiration in much the same way as he did the Old Testament prophets. Christians of the first century and afterward have shared the same reverence of the Scriptures as was held by the Jewish people about the Old Testament. Those two Apostles were Peter and Paul.

Peter's View of the New Testament as Scripture

In one short passage Peter makes clear his conviction that both his and Paul's writings are Scripture:

> 2 Peter 3:15 *Count the patience of our Lord as salvation, just as our beloved brother Paul* ***also*** *wrote to you according to the wisdom* ***given him****, 16 as he does in* ***all*** *his letters when he speaks in them of these matters. There are some things in them that are hard to understand, which the ignorant and unstable twist to their own destruction, as they do* ***the other Scriptures****. 17 You therefore, beloved, knowing this beforehand, take care that you are not carried away with the error of lawless people and lose your own stability.*

The words "also wrote" imply that Peter and Paul had both written to these readers, and with a wisdom that was given to them and was not their own. The expression means that their content was not from themselves nor was the authority from each other. The only such shared source for Apostles would have to be the Christ himself, who is "head over everything for the church" (Ephesians 1:22). The nature of these multiple letters (v.16) is therefore authoritative.

The remaining words in the sentence of verse 16 make it even clearer that his letters are authoritative. They say "as they [the twisters of Paul's letters] do the **other Scriptures**." Peter implies that Paul's letters are not just his own as an Apostle, but that they are "Scripture."

Why is the authoritative nature of these letters stressed? Verse 17 makes it clear that the spiritual "stability" of the readers' salvation is at stake. The larger context of the passage makes their situation all the more critical that they heed these Apostolic words as Scripture. The Lord is coming in judgment, and they must be ready, "diligent to be found by him without spot or blemish, and at peace" (2 Peter 3:14).

Now that it can be seen that Peter saw Paul's writings as Scripture (as well as his own), let us take a look at how Paul's letters share the same assumption of inspiration.

Paul's View of the New Testament as Scripture

In 2 Timothy 3:16, Paul said that "All Scripture is breathed out by God." Those familiar with the wording of the King James Version of the Bible are accustomed to seeing it as, "All Scripture is given by inspiration of God." God, by the agency of the Holy Spirit, "breathed out" his truth through the minds and personalities of such people as the Apostles. What was breathed out by God was infused into Scripture. But the full context of the passage gives us a still greater understanding of the idea of the Scriptures being inspired by God.

> 2 Timothy 3:14 *But as for you, continue in what you have learned and have firmly believed, knowing from whom you learned it 15 and how from childhood you have been acquainted with* ***the sacred writings****, which are able to make you* ***wise for salvation through faith in Christ Jesus****. 16* ***All Scripture is breathed out by God*** *and profitable for teaching, for reproof, for correction, and for training in righteousness, 17 that the man of God may be complete, equipped for every good work.*

In writing these words to Timothy, Paul confirms what has already been shown. The Old Testament is Scripture to Jew and Christian alike. Paul tells Timothy to stay strong in what he was taught from youth about the "sacred writings" because they make us "wise for salvation through faith in Christ Jesus." The key difference between Christians and the Jewish people is whether or not those sacred writings point to Jesus.

Less clear to some is whether or not the words "all Scripture," in verse 16, must be applied to the Old Testament alone. It has already been shown that Paul and Peter were conscious of the fact that the letters they wrote were Scripture, inspired and authoritative for the growing churches they addressed. Remember, the whole idea of the Old Testament prophecies was that a Messiah was to come and through him a New Covenant would be established. The New Covenant, by definition, means a new and more direct way to relate to God. The Old Testament alone could not take into account what is necessary for the care of Christians and the churches. That would be because it did not take into account the new reality of the Holy Spirit living in the Christian. Only with the Spirit in us could the Christian "be complete, equipped for every good work."

Found in 2 Timothy 4:1-5 is still more insight into what Paul meant by "all Scripture."

> 2 Timothy 4:1 *I* ***charge you*** *in the presence of God and of Christ Jesus, who is to judge the living and the dead, and by his appearing and his kingdom: 2* ***preach the word;*** *be ready in season and out of season; reprove, rebuke, and exhort, with complete patience and teaching. 3 For the time is coming when people will not endure* ***sound teaching,*** *but having itching ears they will accumulate for themselves teachers to suit their own passions, 4 and will turn away from listening to* ***the truth*** *and wander off into myths. 5 As for you, always be sober-minded, endure suffering, do* ***the work of an evangelist,*** *fulfill your ministry.*

"Preach the word" must include all the teaching of Jesus and the other inspired writings of the Apostles in order to be the "sound teaching" of verse three. "Doing the work of an evangelist" is to set churches in order as determined by God's will shown to the Apostles (Titus 1:5). That concept is certainly not from the Old Testament. In the Greek text definite articles are used indicating a specific word or work for "*the* word" and "*the* work of an evangelist." Both have very specific meanings. To be charged by Paul to preach *the* word had to have included the New Testament also. To be charged by Paul to do *the* work of an evangelist had to include building churches.

In summary, when Paul said that "all Scripture" is inspired of God he did not merely mean the Old Testament Scriptures of the Jewish people. We know that Paul had to include the inspired Apostolic writings also.

Paul's Apostolic Responsibility

At his ascension into heaven, Jesus was very clear about his authority and by extension about that of those receiving his "great commission," as it has come to be known. In the broad sense the great commission is spoken to all of us who read it, but a special responsibility fell to those who were to launch the church.

> Matthew 28:19 *And Jesus came and said to them, "**All authority in heaven and on earth** has been given to me. 19 Go therefore and **make disciples** of all nations, baptizing them in the name of the Father and of the Son and of the Holy Spirit, 20 **teaching them to observe all that I have commanded you**. And behold, I am with you always, to the end of the age.*

The stress is upon the continuity of the teachings of Jesus. The Apostolic band would be the foundational people in determining this truth, then teaching it to others, who would then teach others to obey it as well. The first act of anyone who decides to follow Jesus is to be baptized (immersed under water). Once baptized his followers are obligated to obey and to teach others to do so also.

Paul reflects the great commission of Jesus to teach others in his letter to the Ephesians.

> Ephesians 2:18 *"For through him we both* [Jew and Gentile] *have access in one Spirit to the Father. 19 So then you are no longer strangers and aliens, 4 but you are fellow citizens with the saints and members of the household of God, 20 built on the **foundation of the Apostles and prophets**, Christ Jesus himself being the **cornerstone**, 21 in whom the whole structure, being joined together, grows into a holy temple in the Lord."*

The analogy is that Christ is the chief cornerstone, and the Apostles were the other foundation stones. On this foundation would rest everything for which the church would be erected, as each Christian became a building block in this spiritual temple.

In I Corinthians 14:37, we find a direct use of his authority, as the final test of anyone else's credibility.

> I Corinthians 14:37 *If anyone thinks that he is a prophet, or spiritual, he should acknowledge that the things I am writing to you are a **command of the Lord**.*

Because Paul understood his role to include teaching Christians to obey from the heart, he was reluctant to "pull rank" in the churches and just command anything. He wanted Christians to operate in the spirit and from the heart. For him to remind anyone of his authority under Christ would require clear situations. This special letter to Philemon addresses one of those situations.

Paul wants a newly converted Christian brother named Onesimus to be freed from slavery by his owner, Philemon:

> Philemon 8 *Accordingly, though I am bold enough in Christ to* ***command you to do what is required****, 9 yet for love's sake I* ***prefer*** *to appeal to you—I, Paul, an old man and now a prisoner also for Christ Jesus—10 I appeal to you for my child [in the faith], Onesimus, whose father I became in my imprisonment. 11 (Formerly he was useless to you, but now he is indeed useful to you and to me.) 12 I am sending him back to you, sending my very heart. 13 I would have been glad to keep him with me, in order that he might serve me on your behalf during my imprisonment for the gospel, 14 but I preferred to do nothing without your consent in order that your goodness might not be by compulsion but of your own accord. 15 For this perhaps is why he was parted from you for a while, that you might have him back forever, 16 no longer as a bondservant but more than a bondservant, as a beloved brother—especially to me, but how much more to you, both in the flesh and in the Lord.*
>
> *17 So if you consider me your partner, receive him as you would receive me. 18 If* ***he has wronged you*** *at all, or* ***owes you*** *anything, charge that to my account. 19 I, Paul, write this with my own hand: I will repay it—to say nothing of your owing me even your own self. 20 Yes, brother, I want some benefit from you in the Lord. Refresh my heart in Christ. 21 Confident of* ***your obedience****, I write to you, knowing that you will do even* ***more than I say****. 22 At the same time, prepare a guest room for me, for I am hoping that through your prayers I will be graciously given to you.*

Note the unmistakable pastoral tone of Paul's words as he appeals to the heart and conscience of Philemon. It is strong moral and ethical leverage used by Paul. From verse 17 onward, Paul deals with any objection, even the financial impact of freeing Onesimus. This is a grand illustration of the nature of servant leadership that should prevail in God's church.

In other places in the New Testament Paul uses his authority in still stronger terms. For example, he tells the Corinthians that they must discipline a man for having his father's wife.

> I Corinthians 5:3 *For though absent in body, I am present in spirit; and as if present,* ***I have already pronounced judgment*** *on the one*

> *who did such a thing. 4 When you are assembled in the name of the Lord Jesus and my spirit is present, with the power of our Lord Jesus, 5* ***you are to deliver this man to Satan*** *for the destruction of the flesh,* ***so that his spirit may be saved in the day of the Lord.***

It is clear that Paul is presuming the authority to judge this situation, even when he is physically absent. The Corinthians are told to "deliver this man to Satan," meaning put him out of the church in hopes that his situation of loss might bring him to his senses. The goal is that he "be saved in the day of the Lord."
In summary, Paul's Apostolic responsibility was a reality that Paul used in pastoral matters. What he says in his letters is Scripture and to be received as such by churches and Christians alike.

Jesus as God's Best and Last Word

> *Hebrews 1:1* ***Long ago****, at many* ***times*** *and in many* ***ways****, God spoke to our fathers by the prophets, 2* ***but in these last days he has spoken to us by his Son****, whom he appointed the heir of all things, through whom also he created the world. 3 He is the radiance of the glory of God and the exact imprint of his nature, and he upholds the universe by the word of his power. After making purification for sins, he sat down at the right hand of the Majesty on high...*

Hebrews 1:1-3 covers a vast truth not easily known without a knowledge of the Old Testament. It says, *"Long ago, at many times and in many ways, God spoke to our fathers by the prophets."* The words "long ago" take in a period of time that starts with creation itself and stops with the book of Malachi, the last inspired prophet of the Old Testament Scriptures, 400 years before the coming of Christ.

"Many times" certainly include the earliest period of innocence in the Garden of Eden. It would also include the period of the great flood, the bondage and exodus, the wilderness wandering, the early united kingdom, the period of the divided kingdom, the exiles, and the restoration of the nation and temple under Ezra and Nehemiah.

"Many ways" certainly include God speaking vocally and face-to-face to such early patriarchs as Abraham, Isaac, and Jacob, and certainly to Moses. Other ways were through angels, through a burning bush when he called Moses, and perhaps most intriguing of all, he even spoke through a donkey.

But the real point of this citation is made in Hebrews 1:2. *"But in these last days he has spoken to us by his Son."* With the coming of these "last days," the time of the New Covenant, comes the last and greatest revelation from God: his very Son.

> *Hebrews 1:2b-3 ...his Son, whom he appointed the* ***heir*** *of all things, through whom also he created the world. 3 He is the* ***radiance*** *of the glory of God and the* ***exact imprint*** *of his nature, and he upholds the universe by the word of his power.*

His son is said here to be the "radiance of God's glory." The radiance of anything is usually considered a part of the object, or an extension of the object that is doing the radiating. The idea here is that the Father and the Son hold that special inseparable type of connection. When it is said that Jesus is the "exact imprint of his nature" another inseparable relationship is in mind. A stamp or mold is designed to duplicate its exact form on the material stamped. In the case of a coin, the mold or die stamps its very imprint on the metal. The resulting coin is the visible expression of the hidden form in the mold or die that stamps it. So it is that Jesus is the visible expression of the otherwise hidden image of God. Indeed, the incarnation of the son into the flesh of Jesus was a special event to reflect the character and person of God as he relates to us. We are to allow ourselves to be formed closer and closer into that mold, model, or image.

Further, this Son, Jesus, still "upholds the universe by the Word of his power." Just as nothing was made without him (John 1:1-3), nothing continues without him either. We don't understand just how that works, but the Scriptures simply tell us that it is so. Most important of all is that Jesus "after making purification for sins, ... sat down at the right hand of the Majesty on high" (Hebrews 1:3b; 8:10). After his time in flesh, making known the will and ways of God, and paying the purification price for our sins, he reassumed his eternal position with God at his "right hand." What a powerful image of that relationship!

Hebrews 2:1-4 takes this high, eternal, and divine nature of Christ to a logical conclusion. It shows the finality and sobriety with which this new revelation is to be viewed. It is clearly pointed out that if God severely judged those who received the older, even inferior revelation, in the Old Testament, he must do so even more when it comes to those refusing this final revelation: his son.

> *Hebrews 2:1-4 …we must pay* ***much closer*** *attention to what* ***we*** *have heard, lest we drift away from it. 2 For since the message declared by angels [for example under Moses] proved to be reliable, and every transgression or disobedience received a just retribution, 3 how shall we [since the coming of Jesus] escape if we neglect such a great salvation [based in Christ as the Messiah]?*

The Faith Once and For All Delivered

The finality of this new revelation from God, in Jesus, is put very succinctly by Jude in the short letter which immediately precedes the book of Revelation.

> Jude 3 *Beloved, although I was very eager to write to you about our common salvation, I found it necessary to write appealing to you to* ***contend for the faith that was once for all delivered to the saints****. 4 For certain people have crept in unnoticed who long ago were designated for this condemnation, ungodly people, who* ***pervert the grace of our God*** *into sensuality and* ***deny*** *our only Master and Lord, Jesus Christ.*

Jude was desirous to exhort us in our "common salvation." But the reality of false teaching made him issue a strong warning instead. That warning is just as relevant today as it was then. The message of Jude is simply that salvation by grace does not preclude the necessity of obedience. False teachers were misleading people into thinking that faith doesn't require change and obedience on the part of the Christian. Such false teachers were those "who **pervert** the grace of our God into sensuality and **deny** our only Master and Lord, Jesus Christ."

Most of all for our purpose, note the words "**the** faith **once for all** delivered to the saints." Preceding the word faith is a definite article that exists even in the original Greek – "*the* faith" means a very specific identifiable faith. THAT Faith is the one that was "**once for all** delivered to the saints."

The words "once for all" surely convey that the original message was not intended to change, or evolve. What was taught at the beginning was expected to be the same for all who would believe, no matter how many years in the future that might be.

In his last days before crucifixion Jesus said clearly to Thomas, "I am the way, and the truth, and the life. No one comes to the Father except through me." Surely he shows that God has only one plan of redemption.

In a defense by the Apostles before the religious leaders of Jerusalem, Peter said

> *Acts 4:10 Let it be known to all of you and to all the people of Israel that by the name of Jesus Christ of Nazareth, whom you crucified, whom God raised from the dead—by him this man is standing before you well. 11 This Jesus is the stone that was rejected by you, the builders, which has become the cornerstone. 12 And there is salvation in no one else, for there is* ***no other name under heaven given among men*** *by which we must be saved.*

Paul certainly taught the finality of this gospel message when he spoke of false teachers in Galatia. There is no room in this statement of Paul for future changes in God's message to mankind, even if such a message is said to come from angels.

> Galatians 1:8 *Even if* ***we or an angel from heaven*** *should preach to you a gospel contrary to the one we preached to you, let him be accursed. 9 As we have said before, so now I say again: If anyone is preaching to you a gospel* ***contrary to the one you received****, let him be accursed.*

In one last citation on this subject, we find words in the last book of the New Testament, the book of Revelation. These words echo those found in the Old Testament concerning the Law given to Moses:

> Revelation 22:18 *"I warn* ***everyone*** *who hears the words of the prophecy of this book: if anyone adds to them, God will add to him the plagues described in this book, 19 and if* ***anyone*** *takes away from the words of the book of this prophecy, God will take away his share in the tree of life and in the holy city, which are described in this book."*

In Summary

Jesus was clear about how authoritatively he viewed the Old Testament Scriptures and that they would not pass away until the very end. Jesus was also clear that the Holy Spirit would make the very words of God clear to the Apostles. The Apostles themselves were aware of the authority by which they wrote these inspired accounts; therefore, both Peter and Paul are clear about the nature of New Testament inspiration and authority. Many passages teach that Jesus is God's last revelation. Therefore with the death of the Apostles was the close of Scripture for Christians. This message was given once, and for all, to God's people. It was not to be changed, either by adding to it, or by subtracting from it.

My wife's 5th Great-grandmother gave the Bible shown here to her son Edwin Bacon in 1771. It was printed in 1766 and has been passed down through the family.

The Table of Contents lists the Apocryphal books, yet they are not actually published in the text itself.

Discussion

1. How is the written documentation of the New Testament much like that of the Old Testament?

2. Who really "created" the New Testament? Was it the church at large, the Apostles (by inspiration), some council of men centuries later, or some other entity?

3. What is the gospel, or the good news, according to the New Testament? Why is this gospel central to our salvation?

4. How important is the church to Jesus, our Savior? How important should it be to us?

5. What parts of the New Testament are the most important for us to know? How often should we read it?

6. What devotional and study aids might be helpful for Christian growth?

CHAPTER 3:

Sound Standards of History

"...that you may have certainty concerning the things that you have been taught."

Luke 1:4

Three authors in the New Testament are specific in how they use history as a foundation of gospel evidences. They are Paul, Luke, and Peter. Let us look now at how each one appeals to history.

Paul's Use of Historical Content

Paul's first letter to the Corinthians contains a very strong statement of the historical foundation of the gospel. It is dated about 53 A. D. That is 20 years after the generally-accepted resurrection date of 33 A. D.

> I Corinthians 15:1 *Now I would remind you, brothers, of the gospel I preached to you, which you received, in which you stand, 2 and by which you are being saved, if you hold fast to* ***the word*** *I preached to you—unless you believed in vain. 3 For I delivered to you as of* ***first importance*** *what I also received: that Christ died for our sins* ***in accordance with the Scriptures****, 4 that he was buried, that he was raised on the third day* ***in accordance with the Scriptures****, 5 and that he* ***appeared*** *to Cephas, then to the twelve. 6 Then he* ***appeared*** *to more than five hundred brothers at one time,* ***most of whom are still alive****, though some have fallen asleep. 7 Then he* ***appeared*** *to James, then to all the Apostles. 8 Last of all, as to one untimely born, he* ***appeared*** *also to me. 9 For I am the least of the Apostles, unworthy to be called an Apostle, because I persecuted the church of God. 10 But by the grace of God I am what I am, and his grace toward me was not in vain. On the*

> *contrary, I worked harder than any of them, though it was not I, but the grace of God that is with me. 11 Whether then it was I or they, so we preach and so you believed.*

Paul firmly says that certain aspects of the message are of "first importance," that Christ died and was raised.

Concerning this death and resurrection, he states a key point twice, the first being verse three, where he says that "Christ died for our sins, in accordance with the Scriptures." Then, he says in verse four that "he was buried and he was raised on the third day, in accordance with the Scriptures." This repetition of the words "according to the Scriptures," is not an incidental redundancy. It is a clear means of emphasis. The point would be that what was foretold in the Old Testament can now be seen clearly in the life of Jesus. That message was widely made known in preaching and in the written accounts.

But, secondly, Paul notes that many of those who spread this word were witnesses and they were still alive. The list of witnesses in the order given are Cephas (Peter), the twelve, more than five hundred brothers (most of whom still were alive), James, and the Apostles again. That so many witnesses were still alive certainly means that they believed they saw Jesus alive, according to the Scriptures that spoke of his not staying dead, but defeating death for all of us. Skeptics of that time faced solid testimonies from many.

Only at the end of this section of evidence does Paul bring in his own testimony and experience as an Apostle. In verses 10-11 we see that Paul worked hard by the grace of God to spread the message that Christ has risen. He put it as, "Whether then it was I or they, so we preach and so you believed."

Paul's appeal to "according to the Scriptures," flowing from the Old Testament, is not stated only here to the Corinthians. In one form or another it is in all of his letters. Another reference is especially clear as to how Paul bases his presentation of the gospel on the promises in the Old Testament. It is found in the opening of his second letter to Timothy.

> 2 Timothy 2: 8 *Remember Jesus Christ, risen from the dead, the* ***offspring of David****, as preached in* ***my gospel****, 9 for which I am suffering, bound with chains as a criminal. But the word of God is not bound!*

In summary, Paul had a strong view of the Scriptures as historical events, and the news of these events was spread widely by many credible witnesses who were still alive. These historical gospel events were said to be rooted in historical promises made in the Old Testament Scriptures to and through historic figures of the Jewish faith: Abraham, Isaac, Jacob, and even David. Now, let us look at Luke's sensitivity to this history.

Luke's Use of Historical Content

Both the Gospel of Luke and the book of Acts were written by Luke with Apostolic help and approval. More than any other writer of the New Testament, Luke is careful to preface his works by reference to his use of the best possible sources and authority.

Luke's Gospel

Luke's Gospel was written as early as the middle fifties, during his journeys with Paul. This would be but 20 years after the ascension of Jesus from the Mount of Olives. Care for integrity is shown in his preface to that gospel.

> Luke 1:1 *Inasmuch as* ***many have undertaken to compile a narrative*** *of the things that have been accomplished among us, 2 just as those who* ***from the beginning were eyewitnesses and ministers of the word*** *have delivered them to us, 3 it seemed good to me also, having followed all things closely for some time past, to write an* ***orderly account*** *for you, most excellent Theophilus, 4 that you may have* ***certainty*** *concerning the things you have been taught.*

He first describes a broad number of narratives that had been compiled, then he notes that those who "from the beginning were eyewitnesses and ministers of the word have delivered them [the narratives] to us" (v.2). Written accounts were plentiful at the time. Further, those who were Apostles and eyewitnesses "delivered them [the accounts] to us."

To be credible, an Apostle had to have witnessed both the ministry and the resurrection of Jesus (Acts 1:21-22). However, please note that those who delivered the narratives to Luke and his company were more than witnesses of the ministry and resurrection of Jesus. They were also "ministers of the word." In other words, those who brought these narratives were more than qualified to judge their authenticity; they even obeyed them to the point of preaching and teaching them to the world. What better foundation could there have been than for the sources to be Apostolic and for the final product to be subject to review by Paul, Luke's travel companion. This must have been what Luke had in mind by saying he wanted "to write an orderly account ... that you may have certainty."

Luke's Acts of the Apostles

In the beginning of what is popularly called "The Book of Acts," there is still another preface about evidences, one similar to the one in Luke's gospel.

> Acts 1:1 *In the first book, O Theophilus, I have dealt with all that Jesus began to do and teach, 2 until the day when he was taken up, after he had given commands through the Holy Spirit to the Apostles whom he had chosen. 3* ***He presented himself alive to them*** *after his suffering by* ***many proofs****, appearing to them during forty days and speaking about the kingdom of God.*

The appearance of Jesus alive, after suffering according to the Scriptures, is crucial to the authenticity of the message. He is said to have done so "by many proofs, appearing to them during forty days." Further yet, Jesus spent this time "speaking about the kingdom of God."

Peter's Historical Appeal

In what follows, Peter shows an awareness that he is writing historically. Because he knows he is about to leave this life for his reward, he wants to remind readers that he was there, a witness to the life, death, and resurrection of Jesus. Peter shows the authenticity of his message when he speaks as only a witness can:

> II Peter 1:14 *... I think it right, as long as I am in this body, to stir you up by way of reminder, 14 since I know that the putting off of my body will be soon, as our Lord Jesus Christ made clear to me. 15 And I will make every effort so that after my departure you may be able at* ***any time*** *to recall these things.*

> *16 For we did **not follow cleverly devised myths** when we made known to you the power and coming of our Lord Jesus Christ, but we were **eyewitnesses** of his majesty. 17 For when he received honor and glory from God the Father, and the voice was borne to him by the Majestic Glory, "This is my belove Son, with whom I am well pleased," 18 **we ourselves heard this very voice** borne from heaven, for **we were with him** on the holy mountain. 19 And **we have the prophetic word more fully confirmed**, to which you will do well to pay attention as to a lamp shining in a dark place, ... 20 knowing this first of all, that **no prophecy of Scripture** comes from someone's own interpretation. 21 For no prophecy was ever produced by the will of man, but **men spoke from God as they were carried along by the Holy Spirit**.*

Verses 16-18 show that Peter roots their faith in a real relationship with a Jesus who existed in bodily form. Jesus is no cleverly devised myth to Peter, who not only knew him so personally, but also Peter was an eyewitness to many of the events about which he wrote. "*We* [includes the other Apostles] heard the voice from heaven," and "*we* were with him on the holy mountain." This language seems chosen to support a Messiah who came in the flesh. That is why Verse 19 contains the wording about the "prophetic word." The inspiration of those prophesies comes from inspired men "carried along by the Holy Spirit."

The Unified Theme of the Bible

Though written by diverse authors from diverse locations in diverse periods of time the Bible is uniquely unified.

Its 66 books were written in three different languages, by 36 or more different authors, over a period of 1,600 years. Just think, people who did not know one another, or even speak the same language, were led by God to write the different parts of what has come to be our Bible. These authors were often not only centuries apart, but in many cases they were of greatly different backgrounds.

The Bible was not written by professional religious people whose job it was to know all about theology and religion. In some cases such people did have a part in it, but for the most part the Bible was written primarily by people who were not professional priests, nor ministers as we know them.

Moses, for example, was not a priest, but he was an educated man, "educated in all the wisdom of the Egyptians" (Acts 7:22). He became, though, a mighty man of God, responsible for the first five books of the Bible. Another educated author used by God in the writing of the Bible was Luke. Luke wrote two books of the New Testament portion of the Bible: Luke and Acts. From all that can be learned about him, we know that he was probably both a physician and a historian. Paul was another great human instrument in writing the Bible. He wrote as many as 14 of the books (actually letters, in his case) of the New Testament. He was a student who was tutored by a scholar of great reputation (Acts 22:3).

Men of great learning were not the only types used by God in the writing of the Bible, however. David was a shepherd boy raised up by God to be King of Israel as well as an instrument in the transmission of His Holy Word. Amos was a shepherd called by God to be a prophet and a writer of His Word (Amos 1:1). Additionally, we must not forget the unlikely assembly of 12 people whom Jesus called. Some were fishermen, at least one was a tax collector.

But the uniqueness of the Bible can be seen in other ways. There are different styles of writing in the Bible: genealogy, history, religious law, prophecy, proverb, letter, gospel, song, and even a sexually suggestive poem.

Stop and think about it. Wouldn't you think with so many authors, writing in different languages, in different styles of writing, and over so many hundreds of years, that there would be many mistakes and contradictions?

That the Bible has such a unified history is evidence of God's supervision of its development. A further note on the reliability of God's inspired word might be the fact that the Bible does not "whitewash" uncomfortable truths about Bible characters. As you become more familiar with the Bible, you will see that great Bible characters have their shortcomings and sins as well as their great strengths. Such is the honesty of the Bible.

However, over time the Bible has passed through human hands that have not been inspired. Various copies of Bible manuscripts that date many years after the writings of the Apostles have been found to have some copying errors in them; translational errors occur occasionally also. These are errors however that do not reflect on the original writing of the Bible because they are errors by men at later dates than the dates of the original authors. If your Bible has footnotes in it, you will see that these notes sometimes call attention to what some alternate manuscript reading might have. But in no case do alternate readings change the message.

In summary, the books written by Paul, Luke, and Peter give us much insight into just how they were inspired in their content and oversight. Mention of the early sources as being eyewitnesses and ministers of the word, should make us realize how historical and authentic the written sources are. In our next chapter, the compilation of early Christian writings compare well to the compilation of recent historical accounts.

For quite a while the Muratorian Fragment was held to be the oldest existing fragment of a NT document, dating to 125 A. D. But now there is a strong case for still older pieces of New Testament literature, dated to be around 60 A. D. One is known as the Magdalen Papyrus. The other is simply a fragment of Mark from the Dead Sea Scrolls, cave #7. A book about this research is *Eyewitness to Jesus*, Thiede and D'Ancona, 1996. Dr. Carsten Thiede was a German Biblical scholar and an expert in Papyrology, old documents written on papyrus. Among other things Dr. Thiede examined the material under his sophisticated "epiflourescent confocal laser scanning microscope. Unlike other microscopes this can differentiate between twenty separate micrometer layers of a papyrus manuscript" (p.60). These and other results demonstrate the believability of the Bible when it says that these accounts were actually written that early.

Discussion

1. Which of the three Apostles cited in this chapter is the most compelling in a concern for history, and why?

2. If you were to recommend only one of the four gospels for reading by someone who is curious about the faith, which one would it be, and why?

3. Why is the Christian salvation claim so exclusive?

4. What is a simple way to share the nature of the Bible with friends who do not know about it?

CHAPTER 4:

Documenting History

"Get it all on record now - get the films - get the witnesses - because somewhere down the track of history some b --- will get up and say that this never happened."

General and President, Dwight Eisenhower,
upon liberating the Dacau concentration camp.

I had the privilege of serving our country in the United States Army from 1968 to 1971. Much of that duty was spent in what was then West Germany. I was stationed in the city of Nurnberg, know to most Americans as the place of the Nurnberg Trials after World War II. Nurnberg was also a center of major Nazi Party activities and was planned by their leaders to play a key part in the long range destiny of Hitler's Third Reich.

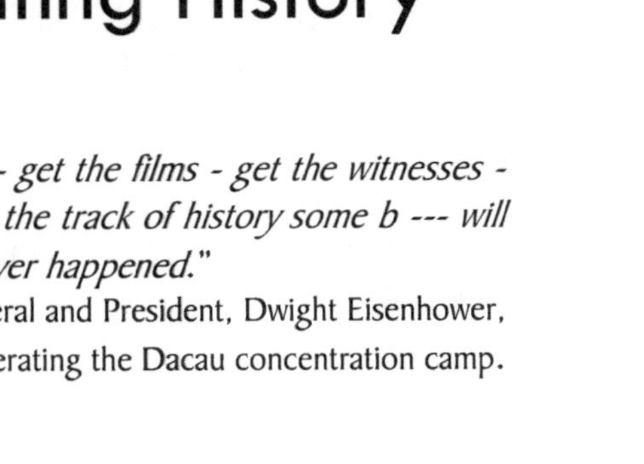

George and RuthAnn Hall stand in front of the former SS barracks near the Nazi Congress Hall. It was first used as a U.S. Army base for many years. Ironically, it is now a processing center for immigrants to Germany.

The barracks in which I was housed for part of that time was a former headquarters and barracks facility for Hitler's elite SS troops. That facility was but one piece in a larger planned complex for the future direction of Germany. Next to the barracks was the famous parade field for many of Hitler's party rallies. Next to it, was the planned huge Congress Hall patterned after the Roman Coliseum, only larger. Still other grand buildings were planned as well. Those plans were ultimately stopped by the intensity of their war efforts, allied bombing, and Germany's unconditional surrender on May 7, 1945.

By the time I served in Nurnberg, the city was mostly rebuilt and thriving, only 25 years after the war. The actual buildings of many historic places remain, though many had to be rebuilt, including much of the old castle and walled city. Those structures that did not require rebuilding were the parade field and the Congress Hall. The SS barracks seen above was renamed Merrill Barracks which then housed various U. S. Army military units through the Cold War. That this history is true cannot be doubted.

In 1970, many people who knew the war firsthand still lived. Between 1970 and 2007, I met many Americans who served in the European Theatre of WWII. A man who had been part of liberating Dachau spent months in a psychiatric ward dealing with his trauma. Others I had met were Holocaust survivors, and sometimes they would share but a glimpse of their experience. Such heroic stories make us grateful for the courage of others who faced down the evil. Most of the individuals who shared some of their story would not consider writing their accounts. It was too painful, and they felt inadequate. That this history is true cannot be doubted.

In 2007, I was able to tour this area again, along with my wife and oldest daughter, Hannah. Much had happened in subsequent German history, the most significant being the fall of the Berlin Wall and the whole Iron Curtain. By this time there were museums established in Nurnberg that freely dealt with the dark shadow of the former Nazi regime. People were more and more conscious of the fact that future generations needed to have this history preserved for the record. Even in Washington, DC, a Holocaust museum had been created by this time.

Of special importance to me was being able to travel freely by train from Nurnberg to Berlin. Its path went through the Hof Corridor, once an area where Soviet and NATO armor units were face-to-face across the border from each other practicing for possible combat during the Cold War. What a difference a few years can make! But the lessons of those years should always be remembered and taught. That is the job of history.

The upper left picture shows Hitler and his construction people with a scale model of the Congress Hall. The Hall can be seen in the background, still under construction. The upper right picture shows the Congress Hall while wrapped with scaffolding. The lower right shows the nearby Zeppelin Field, the site of the many famous Nazi rallies. These three pictures are used by the kind permission of the Documentation Center, now housed in the former Congress Hall (Dokumentationszentrum Reichsparteitagsgelände). The lower left picture however is a more modern one. It is a contemporary aerial view, even showing tour buses in the parking lot. The lower left picture by Nico Hoffman is used by permission and can be found at *de.wikipedia.org/wiki/Datei:Nuremberg_Aerial_Kongresshalle*

While traveling in Poland, we were able to fulfill a dream to visit the site of Auschwitz, once the notorious hub of several concentration camps, but now a memorial.

Main gate of Auschwitz, key Nazi concentration camp in Poland. The infamous sign says "work makes free."

Pictured above is a restored section of the elaborate security fencing at the camp. Layers of fencing, electrification, dogs, and armed guards secured the inmates against escape.

Some thought General Dwight D. Eisenhower was over-reacting when he wanted the concentration camps and atrocities of the Germans to be heavily documented by witnesses, in photography and film. His reason was that he knew some would believe that the stories were exaggerated or even untrue. He also knew that future generations would doubt the facts once real witnesses were no longer here. Mere verbal testimony would be easier to contest by future citizens, researchers, and writers. (Further exploration can be enhanced at http://www.nizkor.org/, especially the tab titled "Deceit & Misrepresentation: The Techniques of Holocaust Denial.")

Technological advances have made the recording of modern history much better through the mediums of photography, sound, and film that now exist. Just having these pictures from German history seems so much more powerful than mere stories that could be written.

But such documentation of the Holocaust has still not stopped many people from doubting the events. They prefer to attribute the evidence to Jewish, capitalistic, and western conspiracies. Many follow the leadership of people such as Mahmoud Ahmadinejad, a former leader in Iran who believes that the Holocaust never happened. It is not convenient for some to accept this history because of their own religious beliefs and hatred of Jewish people.

Ironically, skeptics of the New Testament are not above doing the same thing about the historical value of the Bible books. Let us look at parallels of the New Testament's background with that of the Holocaust.

Contemporary Eyewitnesses

Just as the history of Nazi Germany's Holocaust depends on eyewitness accounts, so too does the history presented in the New Testament. Hundreds of eyewitnesses to the resurrection still existed in the later days of the lives of Luke and Paul. These witnesses could confirm or rebut stories that circulated during their lifetimes. At the writing of the first letter to the Corinthians (about 52 A. D.), Paul confidently stated that most of over 500 witnesses still lived to tell and verify the story (I Corinthians 15:6).

Still more eyewitnesses to the other various miracles and events of the life of Jesus certainly existed. Many more who may not have seen Jesus after the resurrection could confirm many of the other events that happened in and around Jerusalem. After his public execution and burial, many who had witnessed his healings and miracles certainly recalled those events also as they heard the accounts of the resurrection.

Much Written Testimony

Writing and testimony about the Nazi atrocities is voluminous since the war. Most publicized and authoritative was the Nurnberg Trial, where those responsible Nazi leaders were put on trial resulting in weeks and weeks of sworn testimony. Much has been written in the decades since the Holocaust events. Even as late as the 1970s, facts were still coming out about the atrocities. Authoritative histories, and even museums, have come into existence over the following decades.

Similarly, 20 years after the death, burial, and resurrection of Jesus, Luke tells us that many had already undertaken to write accounts of what had happened (Luke 1:1-3). By the end of another 10 years (30 years total) after the resurrection more New Testament letters and the book of Acts had been written. According to Luke and Paul, many of those witnesses were still alive and could recount their experiences and memories at will. They could, and did, confirm believers and challenge skeptics. As that generation of resurrection witnesses began to pass, a more conscientious effort was made to preserve those facts of the faith.

Definitive History

Like the historians of Jesus' day, recent historians made great efforts to preserve the testimonies of Holocaust survivors who would talk or write for the record. Many were still alive over thirty years later and could easily challenge or enrich the existing accounts of the Holocaust. The basic narrative of the history was reinforced however not only by survivors but by recorded and filmed events such as the operation of the concentration camps. The court testimonies of the Nazi leaders have been preserved through transcripts of the Nurnberg Trial.

In a wing of the former incomplete Nazi Congress Hall is now a museum called the Documentation Center. This Center features an architectural glass "shaft of light" that pierces completely through it. The shaft symbolically serves to flood the light of truth on the dark lies perpetrated by the Nazi Party on their nation. By the time of the fall of the Berlin Wall (1989), the darkness of Communism had also come to be pierced by that truth of light and freedom.

As the eyewitnesses to Jesus and the resurrection died, greater appeal was made by Christian leaders to the inspired and reliable record of the first generation of believers. This "canon" (standard) was the accepted truth, the Word, to be passed to future generations. In the case of the New Testament most of those books had been written by around 65 A. D. That is only 30 years after the Resurrection event!

So, the recording of the early facts of the Christian faith compares well to the recording of more recent historical events. Modern readers might not choose to believe, but the historic nature of the writings remains.

The Irony of Photo Evidence

It is easy to wish that photography and other modern tools of history had existed in the first century. But even pictures require a context to have value, and the integrity of any context has to satisfy the scrutiny level of a skeptical viewer.

Take for example this picture of my daughter Hannah standing in Berlin. To her right, near her feet, is the laid bricked marker line of the former Berlin Wall.

A skeptic can dismiss this picture as evidence that the line is authentic by saying it was digitally modified by the "developer" and worthless as having any meaning. To "prove" that this location is where the Berlin Wall once stood, the researcher must consult recorded sources like newspapers, Stasi files(East German state security), and ultimately the testimony of witnesses who may yet be alive. The story yielded by such research must have coherence with what is known within the broader context of German and Cold War history. The skeptic can always choose to ignore the generally accepted story for another version. That is how history works. With that in mind, even photography is only as good as the historical context it is said to reflect.

In summary, the window of time for the eyewitness transmission of history is every bit as short for the New Testament as it is for other historic realities, such as the Holocaust. Skeptics are certainly free to disbelieve the testimonies of the Apostles. But they cannot credibly say that the records themselves are not historical. Even non-Christian historians should be able to admit that a large number of people believed that what they saw was true. The issue is one of accepting the testimony of the Apostles as to whether Jesus was the foretold Messiah of the Old Testament. The history left by the Apostles and early Christians confidently affirm that he has come.

None of these Apostles stood to gain anything by knowingly perpetrating what they ***knew*** to be a lie. Why would they willing die if it had not actually happened? They had nothing to gain. Furthermore the nation within which the resurrection happened had every reason to present and preserve any and all evidence to the contrary. Yet there is no hard evidence to contest Apostolic claims. We can reasonably conclude that the recorded events must truly have occurred.

If we believe that he is the Christ, or Messiah from God, we must also believe his mission: he came to save us from our sins, and restore us to his original design for our lives. Hence we must believe that we are sinners, and therein is the catch. Jesus dying on a cross for sinners is seen as foolishness. We will look at that subject in chapter seven. But comments on still more evidence come next.

Discussion

1. What is a favorite piece of history for you, and how do you know that it is reliable?

2. What lessons about history can be drawn from how the September 11, 2001 terror attacks have been treated by commentators?

3. Do you have old family photographs that remain unmarked as to the names of the people shown, and the date and location of it being taken? How does this show the limits of photographic evidence for others who were not part of the event?

4. What is a favorite museum and what makes it that for you? How do the displays get rotated, and how are they subject to interpretation and even capable of errors?

5. What is an obvious error in history that was later corrected? How did that correction happen?

CHAPTER 5:

Sources from Outside the Bible

> *"Now there was about this time* ***Jesus****, a wise man, if it be lawful to call him a man; for he was a doer of wonderful works..."*
>
> Josephus, Jewish Historian

That Jesus was a living and historical person does not require the Bible alone for evidence. That there was a person named Jesus who was believed by many to be the Christ is an easy point to support. There are several references to Jesus in early historical accounts, and there are still many more accounts of the existence of his followers, the Christians.

Jewish Citations – A very interesting Jewish source is that of Flavius Josephus, a Jewish person who recorded Jewish history for Roman patrons. His clarity in references to Biblical characters is so favorable that it has invited skeptics to dismiss the references; the usual approach by skeptics is to say that Christian believers must have rewritten those passages after the death of Josephus:

> *"Now there was about this time* ***Jesus****, a wise man, if it be lawful to call him a man; for he was a doer of wonderful works, a teacher of such men as receive the truth with pleasure. He drew over to him both many of the Jews and many of the Gentiles.* ***He was Christ.*** *And when Pilate, at the suggestion of the principal men amongst us, had condemned him to the cross, those that loved him at the first did not forsake him; for* ***he appeared to them alive again the third day****; as the divine* ***prophets*** *had foretold these and ten thousand other wonderful things concerning him. And the tribe of* ***Christians****, so named from him, are not extinct at this day."*
> *Antiquities of the Jews,* book 18, chapter 3
>
> ......

> *"And now Caesar, upon hearing the death of Festus, sent Albinus into Judea, as procurator. But the king deprived Joseph of the high priesthood, and bestowed the succession to that dignity on the son of Ananus, who was also himself called Ananus.... Festus was now dead, and Albinus was but upon the road; so he assembled the sanhedrin of judges, and brought before them* ***the brother of Jesus****, who was called* ***Christ****, whose name was* ***James,*** *and some others; and when he had formed an accusation against them as breakers of the law, he delivered them to be stoned." Antiquities of the Jews,* Bk 20, chapter 1.
>
> ......
>
> *"Now some of the Jews thought that the destruction of Herod's army came from God, and that very justly, as a punishment of what he did against* ***John, that was called the Baptist****: for Herod slew him, who was a good man.... Herod, who feared lest the great influence John had over the people might put it into his power and inclination to raise a rebellion.... Accordingly he was sent a prisoner, out of Herod's suspicious temper, to Macherus, the castle I before mentioned, and was there* ***put to death****." Antiquities of the Jews* Bk 18, chapter 5

Roman sources – Tacitus the historian (56-117 A. D.) writes in the Annals (15.44), about 116 AD, a description of the six-day great fire that burned much of Rome in July 64 AD.

> *"Consequently, to get rid of the report, Nero fastened the guilt and inflicted the most exquisite tortures on a class hated for their abominations, called* ***Christians*** *by the populace.* ***Christus****, from whom the name had its origin, suffered* ***the extreme penalty*** *during the reign of Tiberius at the hands of one of our procurators,* ***Pontius Pilatus****, and a most mischievous superstition, thus checked for the moment, again broke out not only in Judæa, the first source of the evil, but even in Rome, where all things hideous and shameful from every part of the world find their centre and become popular. Accordingly, an arrest was first made of all who pleaded guilty; then, upon their information, an* ***immense multitude was convicted****, not so much of the crime of firing the city, as of hatred against mankind."*

In Suetonius (71-132 A.D.) *Lives of the Twelve Caesars*, 25.4, we find the statement, "As the Jews were making constant disturbances at the instigation of Chrestus [Christ], he [Claudius] expelled them from Rome [49 A.D.]."

Suetonius also makes mention of Nero's persecution, in 16.2: "Punishment by Nero was inflicted on the Christians, a class of men given to a new and mischievous superstition."

Mara bar ("son of ") Serapion was a Stoic philosopher from the Roman province of Syria. He is noted for a letter he wrote in Syriac to his son, also named Serapion. Most scholars date it to shortly after AD 73 during the first century. The letter refers to the unjust treatment of "three wise men": the murder of Socrates, the burning of Pythagoras, and the execution of "the wise king" of the Jews. The author explains that in all three cases the wrongdoing resulted in the future punishment of those responsible by God and that when the wise are oppressed, not only does their wisdom triumph in the end, but God punishes their oppressors.

The letter includes no Christian themes and many consider Mara a pagan.... Some see the reference to the execution of the "wise king" of the Jews as an early non-Christian reference to Jesus. Criteria that support the non-Christian origin of the letter include the observation that "king of the Jews" was not a Christian title, and that the letter's premise that Jesus lives on based on the wisdom of his teachings is in contrast to the Christian concept that Jesus continues to live through his resurrection (wikipedia.org/wiki/Mara_bar_Serapion_on_Jesus).

Pliny, governor of Bithynia from 111-113 AD, wrote the following to the Emperor Trajan, followed by a reply.

> *"It is my practice, my lord, to refer to you all matters concerning which I am in doubt. For who can better give guidance to my hesitation or inform my ignorance? I have never participated in* ***trials of Christians****. I therefore do not know what offenses it is the practice to punish or investigate, and to what extent. And I have been not a little hesitant as to whether there should be any* ***distinction*** *on account of age or no difference between the very young and the more mature; whether pardon is to be granted for repentance, or, if a man has once been a Christian, it does him no good to have ceased to be one; whether the name itself, eve without offenses, or only the offenses associated with the name are to be punished.*
>
> *Meanwhile, in the case of those who were* ***denounced to me as Christians****, I have observed the following procedure: I interrogated*

these as to whether they were Christians; those who confessed I interrogated a second and a third time, threatening them with punishment; ***those who persisted I ordered executed.*** *For I had no doubt that, whatever the nature of their creed, stubbornness and inflexible obstinacy surely deserve to be punished. There were others possessed of the same folly; but because they were Roman citizens, I signed an order for them to be transferred to Rome.*

Soon accusations spread, as usually happens, because of the proceedings going on, and several incidents occurred. An anonymous document was published containing the names of many persons. Those who denied that they were or had been Christians, when they invoked the gods in words dictated by me, offered prayer with incense and wine to your image, which I had ordered to be brought for this purpose together with statues of the gods, and moreover cursed Christ--none of which those who are really Christians, it is said, can be forced to do--these I thought should be discharged. Others named by the informer declared that they were Christians, but then denied it, asserting that they had been but had ceased to be, some three years before, others many years, some as much as twenty-five years. They all worshipped your image and the statues of the gods, and cursed Christ.

They asserted, however, that the sum and substance of their fault or error had been that they were accustomed to meet on a fixed day before dawn and sing responsively a hymn to Christ as to a god, and to bind themselves by ***oath, not to some crime****, but* ***not to*** *commit fraud, theft, or adultery, not falsify their trust, nor to refuse to return a trust when called upon to do so. When this was over, it was their custom to depart and to assemble again to partake of food-- but ordinary and innocent food. Even this, they affirmed, they had ceased to do after my edict by which, in accordance with your instructions, I had* ***forbidden political associations****. Accordingly, I judged it all the more necessary to find out what the truth was by torturing two female slaves who were called deaconesses. But I discovered nothing else but depraved, excessive superstition.*

I therefore postponed the investigation and hastened to consult you. For the matter seemed to me to warrant consulting you, especially because of ***the number involved****. For* ***many persons*** *of every age, every rank, and also of both sexes are and will be endangered. For the contagion of this superstition has spread not only to the cities but also to the villages and farms. But* ***it seems***

> ***possible to check and cure it.*** *It is certainly quite clear that the temples, which had been almost deserted, have begun to be frequented, that the established religious rites, long neglected, are being resumed, and that from everywhere sacrificial animals are coming, for which until now very few purchasers could be found. Hence it is easy to imagine what a multitude of people can be reformed if an opportunity for repentance is afforded."*

Trajan replied to Pliny:

> *"You observed proper procedure, my dear Pliny, in sifting the cases of those who had been denounced to you as* ***Christians.*** *For it is not possible to lay down any general rule to serve as a kind of fixed standard. They are not to be sought out; if denounced and proved guilty, they are to be punished, with this reservation, that whoever* ***denies that he is a Christian*** *and really proves it--that is, by worshiping our gods--even though he was under suspicion in the past, shall obtain pardon through repentance. But anonymously posted accusations ought to have no place in any prosecution. For this is both a dangerous kind of precedent and out of keeping with the spirit of our age."*

Semitic sources – Purdue University History Librarian Dr. Lawrence Mykytiuk has a Ph.D. in Semitic studies. He authored a book titled *Identifying Biblical Persons in Northwest Semitic Inscriptions of 1200-539 BCE* in which he documents in outside sources 50 real people from the Old Testament (check www.biblicalarchaeology.org for a 2014 article on the subject.). Archaeology, in general, also confirms the accuracy of many other place names, locations, and references in the Bible.

History Itself

Beyond specific historical citations for first century events by early historical source is the fact that history itself has been dated for centuries around the person of Jesus. Until very recently historical events were dated as B. C. (before Christ) or A. D. (Anno Domini, Latin for "the year of our Lord"). Lately a more secular BCE (Before Common Era) and CE (Common Era) have been used instead. Either way, the very existence of these time markers of history testify to the existence of Jesus, whatever and whoever one may choose to think that he is.

Discussion

1. Is the citation from Josephus above a favorable one? Does he speak as one who considers himself to be a Christian?

2. Is the citation from Tacitus above a favorable one toward Christians?

3. How can the writings of Christians who lived after the death of the eye-witnesses be reliable history?

4. How might the destruction of Jerusalem by the Romans in 70 AD impact the available history of early Christianity?

5. How might the early persecution of Christians affect the availability of accurate written history?

CHAPTER 6:

Lack of Evidence Is Not the Root Issue

"This is the judgment, ... everyone who does wicked things hates the light and does not come into the light lest ***his works should be exposed."***

Jesus, in John 3:19

Too often the real reason people do not accept God's Word has little to do with the level of the evidence therein. It has more to do with 1) perceived and real religious hypocrisy; and, most importantly, 2) not wanting to face our sins and change.

Reason One: Religious Hypocrisy

Hypocrisy by those who call themselves God's people is nothing new. It is a problem as old as humanity itself, and it is addressed in God's Word. In reading the following passage from Romans it is clear that religious people who judge, but do what they judge in others, will be held to stern judgment by God. That is clear.

> Romans 2:1 *Therefore you have no excuse, O man,* ***every one of you who judges****. For in passing judgment on another* ***you condemn yourself****, because you, the judge, practice the very same things. 2 We know that the judgment of God rightly falls on those who practice such things. 3* ***Do you suppose, O man—you who judge those who practice such things and yet do them yourself—that you will escape the judgment of God?*** *4 Or do you presume on the riches of his kindness and forbearance and patience, not knowing that* ***God's kindness is meant to lead you to repentance?*** *5 But*

> *because of* ***your hard and impenitent heart*** *you are storing up wrath for yourself on the day of wrath when* ***God's righteous judgment will be revealed.***

That the Lord has not returned, or that we have not had a bad consequence for our sin, does not mean we are spared from our guilt. God is patient with us, and there will still be a "righteous judgment" for everyone in the future. Everyone who has lived, now lives, or will live, has to face God's judgment. God is Holy. "Without holiness no one will see God" (Hebrews 12:14). God can do nothing else but what is right when that day comes.

Romans 2:21-24 concludes with a special condemnation and shaming of those who are religious, and who even teach others, yet are guilty of what they fault in others. Those in the church have a special responsibility not to cause the name of God to be blasphemed.

> Romans 2:21 ***You then who teach others,*** *do you not teach yourself? While you preach against stealing, do you steal? 22 You who say that one must not commit adultery, do you commit adultery? You who abhor idols, do you rob temples? 23 You who boast in the law dishonor God by breaking the law. 24 For, as it is written,* ***"The name of God is blasphemed among the Gentiles because of you."***

Hypocrites will be subject to a greater judgment because they discourage others from coming to faith.

But because there are hypocrites in the church is no reason for non-Christians to dismiss their own need for Christ. Can anyone say that he or she is acceptable with God because some other people are hypocritical? How is that logical? Why should a person even want to compare himself downstream? Such a person may be better than that person who is the religious hypocrite, but how does being better than some lost person mean that we are not just another lost person ourselves?

The disengaged critic may defend his belief by adding that at least he or she acts in "sincerity" and consistency without the pretense and show of church attendance. But even sincerity is not without its own inconsistencies and sins. Even the best of people sin against others, hold grudges, lose their moral compass, break vows, fall short in their responsibilities to others, and violate various laws. We are usually blind to our behavior. Therefore a good test is just to ask our spouses, children, and others around us what they see in us.

In summary, hypocrisy is a major problem in understanding and believing the gospel. Hypocrisy in others is no excuse for disobedience, whether we attend a church or not.

Reason Two: Facing the Need to Change

Resistance to change can be seen in the action of a chief religious teacher of that time named Nicodemus, a Pharisee. The Pharisees believed they were already obedient to God because they followed the law of Moses; therefore, they had no need for public repentance shown through baptism. They saw no need to change their present ways, to be "born again." Yet John the baptizer (literally, immerser) and Jesus both insisted on it.

Nicodemus went to Jesus by night to challenge the authority of John and Jesus to call for a public proof of repentance and baptism. Only the Messiah could proclaim such a change and with a new kingdom. The Pharisees would only concede that Jesus must be a prophet from God. Therefore they denied the authority of Jesus. Look at how quickly Jesus answers the challenge of Nicodemus in John 3:2-3.

> John 3:1 *Now there was a man of the Pharisees named Nicodemus, a ruler of the Jews. 2 This man came to Jesus* ***by night*** *and said to him, "Rabbi,* ***<u>we</u>*** *know that you are* ***a teacher come from God****, for no one can do these signs that you do unless God is* ***with*** *him." 3 Jesus answered him, "****Truly, truly****, I say to you, unless one is* ***born again*** *he* ***cannot*** *see the kingdom of God.*

Notice that the plural pronoun "we," underlined above, reflects that Nicodemus speaks for a group, probably those Pharisees. From an explanation in Luke 7:29-30 it is seen that Pharisees refused God's purpose for themselves by not being baptized. Others received him gladly.

> Luke 7:29 *When all the people heard this, and the tax collectors too, they declared God just, having been baptized with the baptism of John, 30* ***but the Pharisees*** *and the lawyers [theologians] rejected the purpose of God for themselves,* ***not having been baptized*** *by him.*

From these verses one can see the division that existed between those who received his word gladly, and those Pharisees who should have known enough to recognize his claim as Messiah. Presumably Nicodemus came "by night" to avoid public exposure. The best explanation is that Nicodemus did not come to learn, but to argue. Notice how Nicodemus sarcastically responds to Jesus' explanation of what one must do to enter the kingdom.

> *John 3:4 Nicodemus said to him, "How can a man be born when he is old? Can he enter a second time into his mother's womb and be born?" 5 Jesus answered, "**Truly, truly**, I say to you, unless one is born of **water and the Spirit**, he **cannot** enter the kingdom of God.*

The common understanding of John and Jesus about being "born again" included the need for public repentance and baptism. Caught off guard, Nicodemus tried to be verbally cute with the words of Jesus in verse four. The comment about entering a second time into the womb is hardly sincere. Nicodemus is not an innocent inquirer about the truth of salvation. That sarcastic remark is why Jesus replied with his second round of "truly truly" in verse five above.

In verse 11 below, Jesus comes back with a third "truly, truly" emphasis that has some more plural pronouns of his own. The pronouns "we" and "you" below define the nature and context of this discussion further. The real subject is about the authority of both Jesus and John to call everyone, even religious leaders, to a new birth shown in repentance and baptism.

> *John 3:11 **Truly, truly**, I say to you, **we*** [John and Jesus] *speak of what **we** know, and bear witness to what **we** have **seen**, but **you*** [Pharisees and religious leaders] *do not receive **our testimony*** [That of John and Jesus as per Luke 7:29-30].

The Pharisees pressed for more evidence to believe that Jesus is more than a "teacher come from God." Jesus told them the heavenly things anyway in John 3:12-18. He said clearly that he, the son of man, is descended from heaven itself. Further, he said, "the son of man must be lifted up [foreshadowing the crucifixion], that whoever believes in him may have eternal life." Everything in verses 12-18 underscores the Messianic claim of Jesus. Nicodemus conceded that Jesus was a teacher from God in verse two, but he refused the higher claims of Jesus. With clarity Jesus tells him what the real issue is: It is their hearts.

> *John 3:19 And this is the **judgment**: the light has come into the world, and people loved the darkness rather than the light because their works were evil. 20 For everyone who does wicked things hates the light and does not come to the light, **lest his works should be exposed**.*

The word "judgment" above carries the meaning of a legal court verdict. The ruling (summary judgment) of Jesus is that he has come into the world as the light. Unbelieving Pharisees love the comfort zone of darkness because they resist change. Self-examination is a fearful and painful thing. The reality of light and truth exposes the reality of sin and pretense. The meaning of the whole passage then is that coming to faith would require a confession of guilt.

If some could refuse the words and miracles of Jesus when they were there with him, they will find reasons to do so now. Some love darkness more than the light, and they are convinced that they are just fine the way they are. That is why many refused the insistent preaching of John and Jesus to "repent for the kingdom of God is at hand." The self-righteous refused to be born again, ignoring their personal need for repentance and baptism. It was not because there was not enough evidence already; they just did not want to have their sins and short-comings brought out by the light of truth.

Summary

Excuses to disobey the call for faith and the mandate that one repent and be baptized involve perceived hypocrisies and resistance to change. Unbelief goes deeper than merely a lack of evidence. It involves our sin, and that is discussed in the next chapter.

Discussion

1. If the Apostles and other witnesses knew that Jesus was not raised as they said, would they have died for a known lie, and for no gain to themselves? How does this differ from some religious martyr who dies for a belief that they believe is true (but cannot know if it is actually false)?

2. How do presuppositions and beliefs that we hold affect our thinking in the course of regular living? For example:

a. How does being a Democrat or a Republican affect how we understand political news reporting?
b. How do stereo-types about gender and class affect our opinions?
c. How would being a single or a divorcee affect how we hear a sermon on marriage and divorce?
d. How would being a minority affect how we react to a discussion against our race that we might overhear?
e. How would the fact that we are guilty of various sins affect our receptivity to the message of the gospel when it says we must repent and turn our lives over to God?

3. How can being guilty of sins make us feel judged, even when a person is not judging us?

CHAPTER 7:

Sin Is Our Root Issue

"For the message of the cross is foolishness to those who are perishing, but to us who are being saved it is the power of God."

I Corinthians 1:18

It is not enough merely to believe that the Bible is credible as history. We must understand what the Bible says about God himself and about our human condition. Let us look first at what the Bible teaches about God.

A Holy God Created Us and Loves Us

The very nature of God is perfect holiness, justice, and love. These are inseparable qualities in God that cannot be diluted. In the following passage God proclaims that he is holy:

> Leviticus 11:44-45 *You shall therefore **be holy, for I am holy**.*

This holy God created us to be like him:

> Genesis 1:1 *In the beginning, **God created** the heavens and the earth. 2 The earth was without form and void, and darkness was over the face of the deep. And the **Spirit of God** was hovering over the face of the waters. … 26 Then God said, 'Let **us** make man in **our image**, after **our likeness**. And let them have dominion over the fish of the sea and over the birds of the heavens and over the livestock and over all the earth and over every creeping thing that creeps on the earth.' 27 So God created man in his own image, in*

> *the image of God he created him;* ***male and female he created them****.*

In the New Testament gospel of John we find corroboration for this creation story when John puts the same story in a bigger context, with more defined content:

> John 1:1 *In the beginning was the* ***Word****, and* ***the Word was with God****, and* ***the Word was God****. 2 He was in the beginning with God. 3 All things were made through him, and without him was not any thing made that was made. … 14 And* ***the Word became flesh and dwelt among us****, and we have seen his glory, glory as of the only Son from the Father, full of grace and truth.*

Jesus is the "the Word." Jesus is God's divine counterpart in creation; he is God's love manifested in human form. John 3:16 tells us the motive for this divine intervention; God so loved the world!

> John 3:16-17 *God so loved the world, that he gave his only Son, that whoever believes in him should not perish but have eternal life. 17 For God did not send his Son into the world to condemn the world, but in order that the world might be saved through him.*

Our Human Condition: Sin Ruined the Relationship

The word for our failure to match God's holiness is the simple word *sin.* In its root meaning the word for sin means to miss the mark. Some dictionaries of the Greek language mention this word as used in connection with archery. When an archer misses his intended target he is said to have sinned. That is just our standing when we are measured against God's holiness: we have missed the mark. That is the meaning when people are spoken of in the Bible as being sinners. We are not totally bad or evil. But we do fall so short of God's character that we must be redeemed from our sin to enter his fellowship.

Sin is not mere mistakes, lapses of judgment, bad circumstances, or a poor upbringing. Sin is something about our human nature that keeps us from loving God and others as we should. Sin is an attitude that falls short of the balance of holiness, justice, and love that is true of God. Sin is an injection of selfishness into our character. We want the traits of holiness, justice, and love on *our* schedule, and by *our* desires. Worse yet, we impose expectations on others that we do not impose on ourselves. Instead of loving people, sin makes us love things and use people. We even misuse God to our own interests.

Sin began in the Garden of Eden when Adam and Eve succumbed to the temptation that they be as God themselves. Here's the account:

> Genesis 3:22 *Then the LORD God said, 'Behold, the man has become like one of us in knowing good and evil. Now, lest he reach out his hand and take also of the tree of life and eat, and live forever—' 23 therefore the LORD God sent him out from the garden of Eden to work the ground from which he was taken.*

The prospect of being "like God" drove them to eat the forbidden fruit. They believed that God was holding back something from them that Satan said they could have. The first temptation was but one expression of what sin involves, and that is distrust of God and his plan for us. God's character dictated that humanity could not be allowed to continue in his presence, meaning they had to be cut off from eternity and "die." The on-going expression of the original sin is that people keep thinking that they can "have it their way," even with God. May the following treatment of the Ten Commandments illustrate how we really do not choose God or his ways.

The Ten Commandments

The standard of morality often adopted on a broad cultural basis is some level of adherence to the Ten Commandments. The commandments are pretty well acknowledged as an ultimate and classic measure of morality and sin. Let us examine all ten with a view to understanding how they might give evidence of our own personal sin.

Some orientation to the whole list is needed first. The first four of these commandments require some extra context and explanation to a modern audience that is not Jewish. The first four commands involve our behaviors toward God much more than one realizes; while the last six involve our relationships to other people.

The Ten Commandments (Exodus 20:3-17)

1 -- You shall have no other gods before me.
2 -- You shall not make for yourself a carved image ... of anything [for worship].
3 -- You shall not take the name of the LORD your God in vain.
4 -- Remember the Sabbath day, to keep it holy.
5 -- Honor your father and your mother.
6 -- You shall not murder.
7 -- You shall not commit adultery.
8 -- You shall not steal.
9 -- You shall not bear false witness...[Do not lie].
10 -- You shall not covet your neighbor's house; ... your neighbor's wife, ...male or female servant, ... ox, ... donkey, or anything that is your neighbor's.

The **first command** assumes there is but the one true God to worship and serve. That means we must not give our allegiance to anyone or anything else. We all follow and serve someone or something. It might be self.

The **second command** accentuates the importance of the first. It precludes using images of that one true and Holy God, or any such representations to enhance worship. God wants an uncluttered personal relationship with each of his people:

> Acts 17: 24-31 *The God who made the world and everything in it, being Lord of heaven and earth, does not live in temples made by man, **25** nor is he served by human hands, as though he needed anything, since he himself gives to all mankind life and breath and everything. **26** And he made from one man ever nation of mankind to live on all the face of the earth, having determined allotted periods and the boundaries of their dwelling place, **27** that they should **seek God, and perhaps feel their way toward him and find him**. ... **29** we ought **not to think** that the divine being is like gold or silver or stone, **an image formed by the art and imagination of man. 30** The times of ignorance God overlooked, but **now** he commands **all people everywhere** to repent, **31** because he has fixed a day on which he will judge the world in righteousness by a man whom he has appointed; and of this he has given assurance to all by raising him from the dead.*

Too often people are willing to let other things fill the spiritual void in their lives, even religious things that may make us feel good. Nothing of a material nature should hinder our growing understanding of who God really is.

The **third command** has to do with insincere, even irreverent, speech about God. Using the name of God and Christ in a cursing fashion clearly is such an example. But it can also extend to such things as using God talk in a flippant way to sound pious when one's life is not holy. Too often those who claim to be a follower of God do not honor him in their behavior.

Strictly speaking, the **fourth command** cannot apply unless one is also a religiously active Jewish person. The Sabbath is the seventh day, and is Saturday, and was given for refocusing on God and one's faith. Christians however, have customarily used the first day of the week, Sunday, to do this. For Christians Sunday is called "The Lord's Day" because it is to be designated for maintaining our focus on Christ and for deepening our faith. That primarily happens in a worship service where the focus is remembering him and drawing closer. Having the opportunity to remember him as commanded in the Lord's Supper goes to the heart of this command. Too many people will justify not going to church because they "can" worship on a golf course or by some other venue. But do they really do it? Further, do those other events tune up our faith or cause us to remember him in the way he commanded? Too many times people erode their faith by infrequent church attendance.

The **fifth command** certainly applies to respectful speech and attitudes toward parents. Unfortunately we interpret this as applying only to obvious respect. Not so obvious is to show respect by treating them as well as we do others in our lives, even in polite and respectful speech. We should pay attention to what they say, tell them the truth, respect their possessions, show gratitude, and speak well of them to others. Even before they earn it, step-parents deserve respectful treatment. We should not lower ourselves to the use of disrespect.

The **sixth command**, "do not murder," is one that most people feel they have safely kept. In Matthew however Jesus takes the command to its spiritual core by saying.

> Matthew 5:21 *You have heard that it was said to those of old, 'You shall not murder; and whoever murders will be liable to judgment.' 22 But I say to you that everyone who is angry with his brother will be liable to judgment; whoever insults his brother will be liable to*

> *the council; and whoever says, 'You fool!' will be liable to the hell of fire. 23 So if you are offering your gift at the altar and there remember that your brother has something against you, 24 leave your gift there before the altar and go. First be reconciled to your brother.*

This commentary by Jesus on the sixth command rightly tells us not to be proud that we have not murdered, *yet*, but to pay attention to the driving passions and lack of forgiveness that cause one to murder. If one pays attention to these internal realities, we will not only "not murder" but will resolve relationships with those around us. There is also the possibility that an abortion has happened. Whatever may have been our thoughts at the time, we need to be at peace with God on this matter also.

The **seventh command** of "do not commit adultery" includes sexual activity with people who are not our spouse. As with the case of the sixth commandment, Jesus puts the focus of this sin on the driving motives and lusts that cause it. Jesus tells us.

> Matthew 5:27 *You have heard that it was said, 'You shall not commit adultery.' 28 But I say to you that everyone who looks at a woman with lustful intent has already committed adultery with her in his heart.*

The real arena for obedience to the seventh command should be what is cultivated or nurtured in one's thoughts. The person who does not feed unfaithfulness and sin in his mind is the one most likely to stay obedient to this command. Otherwise one's "obedience" may only be a matter of a lack of opportunity to sin instead of obeying God.

The **eighth command** of "do not steal" covers more than whatever might be called outright robbery. Simply put the idea is that we should not take what is not ours to take. That might include not paying a bill, not returning something we borrowed, defrauding an employer of time or materials from our place of work.

The **ninth command** of "do not lie" simply means we should be honest by nature. Honesty includes doing one's own homework assignments, fulfilling job responsibilities, making honest legal declaration for such things as taxes and applications for aid or work. It means being honest with friends and enemies alike. Too often we are not even honest with friends because we fear their disapproval. This can include subtle needs for honesty:

A. -- Pretending something is really not as bad as we know it is.
B. -- Blaming others for what we know was really our fault.
C. -- Excusing our bad behavior because of some lame reason we make up, or because we wish to avoid punishment.
D. -- Comparing our behavior to that of one who can be seen as worse, thinking "I'm not as bad as they are."

Right after the passage in Exodus chapter 20 that contains these Ten Commandments, chapter 23 contains more commentary on the nature of God's expectation on honesty. The following is a very helpful, universal insight. Expressed in the passage is the need to avoid giving in to peer pressure when we are expected to tell what we know it is not true:

> Exodus 23:1 *"You shall not spread a false report. You shall not join hands with a wicked man to be a malicious witness. 2 You shall not fall in with the many to do evil, nor shall you bear witness in a lawsuit, siding with the many, so as to pervert justice, 3 nor shall you be partial to a poor man in his lawsuit."*

The **tenth command** of "do not covet" certainly includes envy. We should be content enough with our situations that we do not let our desires make us sin from materialistic or relational motives. But beyond envy is the possibility that our discontents may cause us not to put God first and make an idol of some pursuit driven by our discontents.

But there is a destructive element that we let happen to ourselves. Our discontent can lead to the destruction of our self-esteem. We envy someone else's body size or shape. We envy how handsome or beautiful someone else. We envy how popular someone is with some group of friends. We envy how skilled a person is in some sport or other activity. We envy the money someone else makes, especially if we think we are worth more. We envy how much education someone has. We envy the kind of car driven by another person, or their other other possessions, and on, and on, and on ... Fill in your blanks to realize how much of a bottomless pit this can be. God knows that.

Beyond the very basic breaking of commands by some action, call them sins of commission. There is the element of not doing the good we know to do, when we should do it. These are sins of omission. Avoiding of responsibility and failure to do what we can, is also sin:

> James 4:17 Whoever knows the right thing to do and fails to do it, for him it is sin.

How many of us lament that we did not respond to some need when we could have? How many people admit that they should have done more in some relationship? Often this form of inaction comes from our fears and comfort zones being challenged. The result was a self-preserving action or decision that was damaging.

Evidence of the prevalence of human sin comes from any survey of network television viewing every weekday afternoon. Whatever is not a "soap opera" portraying relationships outside of God's will seems to be a "reality show" of the aftermath of people who live that way. The main diet of programming for those living in the aftermath of selfishness features Maury Povitch, Divorce Court, Jerry Springer, The Test, etc. Then there are all those TV Court shows designed to rule on cases where God has not been honored in the behavior of the show's litigants. That these shows are so very popular must be because so many live in the troubled waters of sinful habits, and not being guided by anyone higher than themselves.

That not many people can take comfort in the 10 commandments starts with the first four. Many people have other gods before him. That is evidenced in how our pursuit of things and pleasure too often comes ahead of setting aside a day to honor him as the fourth command states. Often that is not done because other gods, and their material representations, are the cause of not setting aside time for him. For some, the pursuit of those other things is laced with language that takes his name in vain, breaking the third commandment. When we do not know God's ways, or want to honor them, we conduct our lives in less than God-honoring ways.

The last six commands are broken when the first four are not the foundation for our life. Most people see infractions of those commands more clearly in others, but not in themselves. Jesus called us higher than those commandments when he called us to take account of our inner drives and lust that can result in such things as disrespect, adultery, stealing, murder, lies, and coveting what is not ours. Our communities and our nation benefit greatly when we as a people allow ourselves to be called higher.

In summary, God created us to have a relationship with him, and he loves us, but humans chose and still choose a path of sin. Our sin-littered pathway of life complicates that relationship. The Ten Commandments make us aware of our sinful state. That awareness should drive us toward him for forgiveness. We have nothing to pay God that he does not already have or cannot create for himself, except our free will response of trust. Sin is destructive and corrosive of our well being, and hence that is why God hates it and cannot just overlook it. He wants us to turn to him as a wayward child to a loving parent.

Summarizing the message of this whole book, the message is that, 1) the Bible is God's inspired Word, 2) the Bible record compares well with other how other history is done, and, 3) the message of the Bible is that God really does love us, and he wants us to return to him. More guidance for any reader who wants to know the practical aspect of stepping out in faith is given in Appendix A.

Discussion

1. How much of a factor is our pride to admitting that we are sinners?

2. What kind of sins are people the least likely to want to talk about and why?

3. How do people often lie about the pleasure they find in various sinful forms of behavior?

4. Why might Christians be reluctant to talk about sins among other members of their congregation?

5. How would being open and confessional about our own sins make our lives more attract to those who do not already know Christ?

6. Discuss a Christian personality or celebrity who has made a strong impact on your life and why?

7. Describe the life and witness of a Christian who has had a good impact on your life.

8. How do afternoon TV reality shows reflect human nature?

APPENDIX A:

A Personal Review of "Believing"

"I will show you my faith by my works!"

James 2:18b

It is common for people of a Christian background to place confidence in the fact that they believe all the basics of the faith. They believe that the Bible is God's word and that Jesus came into the world to die on the cross for our sins. They believe that Jesus was raised from the dead and that we must trust in him to go to heaven. But guess what? Satan and the demons believe that too! The reason that Satan and his demons will not go to heaven is that they refuse to actually DO what God says about having faith.

WHAT to Believe: The Gospel

Paul himself describes the gospel he preached as requiring a belief that Jesus was raised from the dead and was the foretold Messiah, descended from David.

> I Corinthians 15:1 *Now I would remind you, brothers, of* ***the gospel I preached to you****, which you received, in which you stand, 2 and by which you are* ***being saved****, if you hold fast to the word I preached to you—unless you believed in vain. 3 For I delivered to you as of* ***first importance*** *what I also received:* ***that Christ died for our sins in accordance with the Scriptures, 4 that he was buried, that he was raised on the third day in accordance with the Scriptures****, 5 and that he appeared to Cephas, then to the twelve.*

The Prophets of the Old Testament foretold the coming of a Messiah who would be unique and infinite. Only one candidate could do what was needed, and this was foretold centuries before the fact. Please study and answer the questions concerning the following references:

A. **Micah 5:2**
1. From what tribe will this Messiah come?
2. When did this Messiah, yet to be born, really have his beginning?

B. **Isaiah 9:6, 7**
1. What will be born according to verse 6?
2. What does verse six list for his names? Which ones reflect a Divine nature?

C. **Isaiah 52:14 and 53:1-12**
1. What must happen to him?
2. Does verse 11 indicate a return to life? Compare this to **Psalm 16:8-11**.

D. Please read **Matthew 26:36-27:54** noting what this infinite and unique Messiah did for us. Imagine the suffering and pain. Try to put yourself in the place of Jesus, imagining what he must have felt.
1. Physically: Abuse (26:67); flogging (27:27-31); and Crucifixion.
2. Emotionally: Pensive of death (26:36-38); lonely and deserted(26:50, 56); falsely charged (26:60); and denied by his close friends, even Peter (26:69-75).

E. **Romans 5:9-11, 18, 19** What does this say about what Jesus did?

In summary, the Bible exists to tell of God's holiness and of his love for his lost creation. He sent his son to die on a cross to pay the price for us to come back to him. But merely believing this is not enough. Even Satan and his demonic hosts believe this. They know it without a doubt. But they refuse to act on what they know.

HOW to Believe -- Obey and Act on What We Know

The goals of all people are really very much alike. All of us want deep love and trust with a special person(s), emotional peace and stability, material security, a sense of self-esteem and significance, respect from others, "life, liberty and the pursuit of happiness," and so on. The difference between the Christian and non-Christian is that the Christian trusts God and chooses to **pursue** these universal goals according to God's Word, and his design for us.

A. **Matthew 22:34-40** How important is our heart toward God?

____

B. **James 2:14-26**

1. How do verse 17 and 26 describe a "faith" that is alone, apart from obedience? ______________________
2. Does verse 19 show that Satan and demons also "believe"?

   ____
3. What does verse 24 say about "faith alone"? Fill in the blank of that verse here, "________ by faith alone."

C. **Luke 13:3** How serious is repentance according to Jesus?

____

D. **Mark 7:17-23** note that many of these sins are attitudinal. Check the ones of which you have been guilty.

evil thoughts	sexual immorality	murder
theft	adultery	greed
envy	malice	slander
deceit	arrogance	lewdness
folly		

E. **Acts 2:38** How does this show that repentance and baptism are conditions to salvation?____

____

F. Have you been immersed in water as a show of repentance for the forgiveness of sins and the gift of the Holy Spirit?

____

Were you of an age to know what you were doing? ____________________

Stepping Out in Faith

Are you willing to repent of sins and be immersed in water as demanded in the Bible? _______ If so, start showing the following actions of faith, with a heart of faith:

1. Begin praying to God by confessing sins, doubts, and needs. Ask him to lead and guide in decisions.

2. Stop the following habits:

   a. Pretending that some sin is not really that bad after all.
   b. Blaming others for attitudes and behaviors that steal our happiness.
   c. Justifying the "rightness" of some sin for us in our particular situation.
   d. Making little deals with God for his "permission" to overlook our sin.
   e. Appealing to the disobedience of someone else to justify our disobedience.
   f. Not accepting blame for the consequences of our own sinful acts, past and present.

3. Deal with problems rather than avoiding them. Talk with spouse and others as needed, etc.

4. Tell the truth. This includes saying what we mean, and planning to do what we say.

5. Love others, in still deeper ways, including how they feel.

6. Attend a Bible teaching church that understands faith in the same way as shown in this chapter.

Consider churches that are not denominational and called something like Christian Church, or Church of Christ. They are part of a network of churches following only the will of Christ for his churches and people.

Believing in Jesus must mean that we live a changed life. We are to be born again his way. Such faith must include a life of active obedience, worship, and positive growth because his spirit lives in us. Once we make God's priorities our own, our needs all fall into place.

APPENDIX B:

The Formation of the Bible as a Book

A very important development in the preservation of old texts was the introduction of book publishing as we know it. Centuries after the Apostles the practice of binding papyrus pages into a book of pages came into dominance. Until that time, Old Testament Scriptures were a collection of scrolls in a box in the temple and various synagogues. Likewise early Christian writings did not exist in book form as we know it. Christians would also use containers of scrolls to be read and studied as a congregation. From the time of Moses to the time of Malachi the claims of a prophet were likely tested by the nature of his claims. The key test of a prophet was the accuracy and consistency of what he said:

> Deuteronomy 18: 21 *And if you say in your heart, 'How may we know the word that the LORD has not spoken?'—22 when a prophet speaks in the name of the LORD, if the word does not come to pass or come true, that is a word that the LORD has not spoken; the prophet has spoken it presumptuously. You need not be afraid of him.*

A modern adage might well be used to sum up this test: "time will tell." Those who claimed divine inspiration had to be tested by whether their message proved to be consistent with what was already known about God and was deemed to be true. Only true prophetic material could be used from the scroll collection known as the Scriptures. The same attitude prevailed among Christians when it came to Apostolic authority and approval for what was authoritative in the early church.

No councils of mere men ruled on what was to be Old Testament Scripture. The late Dr. Samuel Shultz wrote in his Harvard dissertation that there is no actual evidence for the alleged Council of Jamnia at which some have claimed that this determination was made.

As to the New Testament no council could decide what the Apostles had already decided. Early Christians knew this. Authority for this function was given to Apostles alone by Jesus. Currently accepted canon was written or approved by Apostles before they died. Councils like those of Nicea (325 A. D.) and Carthage (397 A D.) did not rule on what Scripture would be, but on what was already understood to be the canon "from their fathers."

Dan Brown's book *The Da Vinci Code* merely picked up what has been a misconception about the purpose of the Council of Nicea. That Council was to work out an understanding of the relationship of the Father, Son, and the Holy Spirit. They could wrestle with theological questions, but they did not have the prerogative of determining for others what would be accepted as the scriptures.

There are presently over 5,500 Greek manuscripts in existence today for the New Testament. The oldest of those are papyrus manuscript fragments that date to the 60s A. D. There are also many old manuscripts that are translations of the Greek ones. These are several thousand in number and support the accuracy of the Greek texts. The collective accuracy of these manuscripts is amazing. Discrepancies among the copies mostly involve spelling, word choices, word order, and such things that can be expected in such a time when all copying was done by hand and for long periods of time.

Classical writings lack the same massive base of documentary evidence as the Bible and especially the New Testament. For example, Plato is dated at 427-347 B. C. and seven old copies exist. The oldest of these copies is said to 900 A. D., 1,200 years later. In the case of Aristotle who lived 384-322 B. C. there are 49 old copies of his works. The oldest of these is dated to 1100 A. D. and is 1,400 years after the fact. As historical sources go, the accuracy of the New Testament should be considered quite favorably.

APPENDIX C:

Thoughts from a Scientist

For those who struggle with reconciling science and the idea of God, I recommend a little book by Michael Guillen, *Can a Smart Person Believe in God?*, Thomas Nelson Publishers, 2004 (cloth, 170 pages). His book blurb reads, "Michael Guillen, Ph.D., is a theoretical physicist and former science correspondent for ABC News. He taught Physics at Harvard for eight years, during which time he won awards for distinguished teaching. He currently serves as President of Spectacular Science Productions. For his work on *Good Morning America*, *World News Tonight*, *Nightline*, and *20/20*, he has won Emmys, Teddys, and an EMMA. Guillen has also published hundreds of articles and reviews in the *New York Times*, *Washington Post*, *USA Today*, *Esquire*, and *Science Digest*." This is a deeply insightful book that is also highly readable.

APPENDIX D:

Thoughts on Other Religions

World religions can roughly be divided into those which reflect an eastern cultural background and those which reflect a western one. Those of a western background find their roots in the Jewish Old Testament and are monotheistic in their view of God. Roughly speaking they are Judaism, Christianity, and Islam. They began near the eastern Mediterranean Sea and moved "westward" compared to "eastward." Each has its own distinctive history and belief system.

Those religions that are called "eastern" generally find their history in Asian areas; hence they are eastern in their geographical relationship to western areas. Religions like Hinduism and Buddhism fit into this category (Confucianism is not revealed religion so much as an ethical system). Eastern religions also tend to have a shared view of God. The emphasis in such practices is often the need to be absorbed into oneness with God via meditation. The goal of such teachings is to transcend this evil world and be absorbed into oneness with spiritual deity (or deities). There is no acceptance of the Christian view of the need for the death, burial, and resurrection of Jesus. In these eastern religious teachings, Jesus is said to be just another great teacher, or revelator from God. The various religious writings of eastern religions are not seen as Scripture in the same way that the Bible is held in the western view. They are seen as holy writings, but contradictory ideas can be held in them, and that is not a problem to the devotees. Revelation sometimes comes through channeling the spirits from the other side. Their goal is harmonizing religions. The Gnostics fit into this tradition because they were finding ways to harmonize the teachings of Jesus and Christianity with their eastern assumptions. Harmonizing Christianity required that they deny distinctive teachings like the incarnation, death, burial, and resurrection of a bodily Jesus.

Of the western religions one historic event clears the field for Jesus to be the last chance of someone being the Old Testament Messiah. The destruction of

Jerusalem and the Jewish temple by the Romans in 70 A.D. forced Jewish people to abandon serious Messianic hopes. When they were not delivered from the Roman legions, they had to decide whether or not they had missed the Messiah and whether that would have been Jesus. With the destruction of Jerusalem came the destruction of such things as genealogical records, making future messianic claims impossible to prove. Most important of all, destruction of the temple made it impossible for future Jewish generations to obey the Mosaic codes that involved the temple. There could be no real nation, and hence no national compliance to the festivals, no atonement and no forgiveness of sins as stipulated in the books of Moses. They were left with Jesus as the last and best candidate for the Messiah foretold in their Scriptures, yet they reject him to this day.

Christians were emboldened in their faith in Jesus by the Roman destruction because it all happened just as Jesus had said it would happen. Josephus the historian records how those Christians in the city fled when the Roman legions pulled their siege away temporarily, giving them the chance to flee according to Luke 21:20-24. The very preservation of a Jewish ethnicity, without its own nation till 1947, may be evidence that God is still not done with Israel until Jesus returns as he foretold (Acts 1:10-11; Romans 11:25).

Islam and the Quran claim Messianic status for Mohammed, but none of the classical expectations of being the Messiah are met. The Jewish genealogy of record requires that the Messiah must descend from Isaac.

> Genesis 21:12 But God said to Abraham, "...through Isaac shall your offspring be named."

Mohammed's line is not based on descending from Isaac, but from his half-brother Ishmael instead. By that concession alone Mohammed is disqualified as being the anticipated "son of David," who is clearly traced to Isaac in the scriptures themselves.

The very nature of the composition of the Quran does not follow that of the Jewish Bible either. The Quran is merely the compiled revelations of Mohammed, whose teachings lack context. For example, the earliest verses contain statements on Christians and Jews that are positive and familial. Toward the end the tone toward these two neighboring religions turns hostile. They are seen as infidels, and subjection and killing are taught, leading radicals to adopt Jihad. The principle for reconciling such contradictions is called Quranic abrogation. This means that where contradictions exist, the latest thing said should be taken as the true view of Mohammed and the Quran.

APPENDIX E:

Gnostic Writings Today

Gnostic "gospels," as they are called, came into being after the time of the original writings of the New Testament. Gnostic teaching was never seen by the early church as representing a credible view of the faith. These were later attempts by people of eastern religious views to harmonize the new Christian faith into their view. Gnostic ideas were an early version of what are today called "new age" beliefs. They taught that Jesus only seemed to appear and that his was not a bodily historical resurrection. They also rejected as "evil" the very God of the Old Testament who foretold the Messiah.

It is no coincidence that two of the best known of such writings are said to come from the Apostles Judas and Thomas. Judas denied and betrayed Christ, then took his own life. How this gospel bearing his name came to be written is not known for sure. But he is portrayed as a hero, with a different message, also attributed to Jesus. It is a message at odds with that of the earliest gospel writings.

One example of the so-called gospel of Thomas is cited here as a sample of why no Christian could accept it as a true inspired writing. The last line of the Gospel of Thomas contains a discussion between Jesus and his disciples about the status of Mary:

> *"Simon Peter said to them, 'let Mary depart from us because women are not worthy of the Life.' Jesus said, 'look, I will lead her so that* ***I will change her to male****, in order that she also may become a* ***living spirit like you males****. For* ***every female who makes herself male will enter the kingdom of heaven****.'"*

Yes, ladies, you read it correctly that your best hope of salvation is in becoming male, and Jesus in this text says of Mary that "I will change her to a male."

APPENDIX F:

Denominational Confusion

The author has written a separate book about denominational confusion and about the need for Christian unity. It is titled: *The Unity of Obedience* (1998). A clear command to the followers of Jesus is that they are to be in unity that the world might believe.

> John 17:**20** *I do not ask for these only, but also for those who will believe in me through their word, 21 that they may all be one, just as you, Father, are in me, and I in you, that they also may be in us,* ***so that the world may believe that you have sent me****. 22 The glory that you have given me I have give to them, that they may be one even as we are one, 23 I in them and you in me, that they may become perfectly one,* ***so that the world may know that you sent me and loved them even as you loved me****.*

It is theoretically possible that the various denominations of Christendom could be united. Of the various models for unity the author shows, a unity of obedience is advocated, hence the title of the book. To the extent that all denominations would try to pattern themselves after the Apostolic church model unity can grow based on God's Word.

Made in United States
Troutdale, OR
11/03/2026

56845122R00052